Advances in Geographical and Environmental Sciences

Series Editors

Yukio Himiyama, Hokkaido University of Education Asahikawa, Japan
Subhash Anand, Department of Geography University of Delhi, New Delhi, India

Advances in Geographical and Environmental Sciences synthesizes series diagnostigation and prognostication of earth environment, incorporating challenging interactive areas within ecological envelope of geosphere, biosphere, hydrosphere, atmosphere and cryosphere. It deals with land use land cover change (LUCC), urbanization, energy flux, land-ocean fluxes, climate, food security, ecohydrology, biodiversity, natural hazards and disasters, human health and their mutual interaction and feedback mechanism in order to contribute towards sustainable future. The geosciences methods range from traditional field techniques and conventional data collection, use of remote sensing and geographical information system, computer aided technique to advance geostatistical and dynamic modeling.

The series integrate past, present and future of geospheric attributes incorporating biophysical and human dimensions in spatio-temporal perspectives. The geosciences, encompassing land-ocean-atmosphere interaction is considered as a vital component in the context of environmental issues, especially in observation and prediction of air and water pollution, global warming and urban heat islands. It is important to communicate the advances in geosciences to increase resilience of society through capacity building for mitigating the impact of natural hazards and disasters. Sustainability of human society depends strongly on the earth environment, and thus the development of geosciences is critical for a better understanding of our living environment, and its sustainable development.

Geoscience also has the responsibility to not confine itself to addressing current problems but it is also developing a framework to address future issues. In order to build a 'Future Earth Model' for understanding and predicting the functioning of the whole climatic system, collaboration of experts in the traditional earth disciplines as well as in ecology, information technology, instrumentation and complex system is essential, through initiatives from human geoscientists. Thus human geoscience is emerging as key policy science for contributing towards sustainability/survivality science together with future earth initiative.

Advances in Geographical and Environmental Sciences series publishes books that contain novel approaches in tackling issues of human geoscience in its broadest sense-books in the series should focus on true progress in a particular area or region. The series includes monographs and edited volumes without any limitations in the page numbers.

Mark Otieno

Sustainable Agroecological Practices in Sub-Saharan Africa in the Face of Climate Change

Springer

Mark Otieno
Department of Water and Agricultural Resource Management
University of Embu
Embu, Kenya

ISSN 2198-3542 ISSN 2198-3550 (electronic)
Advances in Geographical and Environmental Sciences
ISBN 978-3-031-70471-0 ISBN 978-3-031-70472-7 (eBook)
https://doi.org/10.1007/978-3-031-70472-7

This Springer imprint is published by the registered company Springer Nature Switzerland AG
The registered company address is: Gewerbestrasse 11, 6330 Cham, Switzerland

To the students pursuing courses in Agroecology at various levels, the researchers delving into the depths of sustainable agriculture, the lecturers and professors imparting knowledge and shaping minds, the educators spreading awareness and nurturing future leaders, the extension practitioners bridging the gap between theory and practice, and to all the agroecology enthusiasts who understand the profound connection between farming and ecology—this book is dedicated to you.

May your commitment to understanding and implementing sustainable agroecological practices inspire positive change, foster innovation, and cultivate a harmonious relationship between agriculture and our precious ecosystems. Your efforts and dedication are essential in shaping a resilient and sustainable future for all.

Preface

In recent times, the world has faced profound effects of climate change. The consequences, such as temperature fluctuations, heavy rainfall, and intensifying extreme weather conditions, have diverse impacts that significantly affect our global food systems. These impacts are more pronounced in sub-Saharan Africa, a region that relies on agriculture for sustenance and food security. These changes in our atmosphere are influencing drastically the environment, leaving hazardous aftereffects to be mitigated. Climate change is a reality, with drastic consequences that must be addressed in order for us to have a sustainable future. To confront this challenge, strategies and resourcefulness are critical in dealing with unpredictable atmospheric events and impacts. The acknowledgement of the severity of the situation is the first step toward tackling this problem effectively.

The critical significance of agroecological practices in response to the climate crisis plaguing sub-Saharan Africa is highlighted in this book. The intricate link between agriculture, ecosystems, and climate is further highlighted, calling for an overwhelming need to transform farming systems for long-term sustainability. Research, case studies, and practical examples bring to light the potential of agroecology in combating the effects of climate change. Sustainable practices aimed at promoting biodiversity, preserving resources, boosting soil fertility, and curtailing greenhouse gas emissions all provide the reader with a comprehensive glimpse of the topic. To drive home the success stories from different countries in sub-Saharan Africa, the book manages to spark much-needed inspiration among farmers, policymakers, and practitioners to make and sustain change. The vulnerabilities of smallholder farmers in the region must also be acknowledged, thus traditional knowledge in conjunction with scientific advancements must be integrated in order to give farmers the resources, information, and capacity-building opportunities required to deal with changing climatic conditions. Governments, research institutions, civil society organization, and local communities must align to foster better collaboration and innovation to ensure sustainable agricultural development.

It is our firm conviction that agroecological practices that are sustainable can be the answer to achieving food security, and uplifting the resilience of the ecosystem, nurturing the rural development in sub-Saharan Africa. By incorporating climate-friendly agriculture and sustaining ecological balance, a future can be formed where communities grow, ecosystems flourish, and the effects of climate change can be mitigated.

I offer our utmost gratitude to the numerous researchers, authorities, farmers, and organizations who have provided their vital contributions to the knowledge and practices elaborated in this book. Their dedication and promise to continue sustainable agriculture during these hard times of climate change is truly inspiring. We hope this book can serve as a source of wisdom for all those who share this reality of a self-sufficient and prosperous sub-Saharan Africa.

Together, we should start this venture toward sustainable agroecological practices, realizing that the joint efforts will be highly beneficial to the future generations.

Embu, Kenya Mark Otieno

Acknowledgements

I would like to express my sincere gratitude to the Alexander von Humboldt Foundation for providing the financial support through an alumni grant and a research stay in Germany during the summer of 2023, which enabled me to complete this book titled "Sustainable Agroecological Practices in Sub-Saharan Africa in the Face of Climate Change."

Special thanks to Prof. Dr. Ingolf Steffan-Dewenter for hosting me at the Department of Animal Ecology and Tropical Biology, University of Wuerzburg, during the writing process of this book, offering valuable guidance and insights that greatly enriched its content. I am appreciative of Dr. Marcell Peters at the same department for sharing invaluable knowledge and providing exceptional insights on the topics explored in this book. Your expertise and contributions have been instrumental in shaping the narrative.

I extend my heartfelt thanks to Dr. Rosie Trevelyan, the Director of the Tropical Biology Association, for her continuous mentorship and inspiring encouragement, fostering an environment where I could strive for excellence and innovation in my work.

Lastly, I am grateful to the University of Embu and its top management for granting me the necessary research leave to embark on this journey and travel to Germany for the completion of this book. Your support and understanding have been vital to the realization of this project.

Thank you to all those mentioned for your unwavering support and contributions that have made this publication possible.

Contents

About the Author

Mark Otieno is a distinguished scholar and Senior Lecturer at the University of Embu, Kenya, where he also serves as the Director of Research and Extension. With a profound expertise in agroecology, entomology, and environmental biology, Dr. Otieno's academic journey is marked by a series of illustrious achievements and contributions to sustainable agriculture and biodiversity conservation. His work spans multiple continents and integrates interdisciplinary approaches to address critical global challenges.

Dr. Otieno's educational background is robust and diverse. He obtained his Ph.D. in Agroecology from the University of Reading, United Kingdom, in 2011. His doctoral thesis, titled "Assessing the drivers of pollinator and natural pest enemy communities in pigeon pea and field beans crops," laid the foundation for his future research on the relationships between biodiversity and agricultural productivity. Prior to his Ph.D., he completed a Master of Science in Animal Ecology at Kenyatta University, Kenya, where he explored the effects of herbivory on pollinator–flower interactions in Acacia species. His undergraduate education at the same institution culminated in a Bachelor of Education in Biological Sciences, marking the beginning of his journey into ecological and agricultural sciences.

Throughout his career, Dr. Otieno has demonstrated an unwavering commitment to enhancing the knowledge and practices around sustainable agriculture. Since 2022, he has been a Senior Lecturer at the University of Embu, where he leads academic programs in Crop Protection and Range Management. He is deeply involved in teaching and supervising postgraduate research in agroecology, entomology, agroecosystem management, and environmental biology. His dedication to education is reflected in the development of innovative curricula for MSc in Entomology and BSc in Range Management, as well as his exceptional student-lecturer evaluation scores, consistently above 95% from 2014 to 2023.

Dr. Otieno's professional experience transcends academia. He currently coordinates global food security projects supported by the USA government, focusing on Kenya, Burundi, and the Democratic Republic of Congo. His role as the in-country coordinator for the 1890 University of Arkansas at Pine Bluff (UAPB), Center of Excellence for Global Food Security and Defense (CEGFSD)/Capacity Building

Grant (CBG)/ University of Maryland Eastern Shore (UMES) projects underscores his leadership in addressing food security challenges in East Africa through research and community engagement.

From 2013 to 2021, Dr. Otieno served as a Lecturer at the University of Embu, honing his expertise in teaching and research. He also held a prestigious Alexander von Humboldt - Georg Forster Experienced Research Fellowship at the University of Würzburg, Germany, where he researched crop pollination and pest control services with the aim of transferring knowledge and techniques to Kenya. His international research began earlier with a Postdoctoral Research Associate position at The Pennsylvania State University, USA, where he coordinated a United Stated Department of Agriculture-Specialist Crop Research Initiative (USDA-SCRI) funded project on integrated crop pollination for apple and pumpkin crops.

A significant portion of Dr. Otieno's early career was devoted to building research capacity and promoting conservation. He coordinated projects across East and West Africa for the Tropical Biology Association, funded by the European Union-Africa, Carrebean and the Pacific (EU-ACP), focusing on proposal writing, project design, and publication. As a Research Scientist at the University of Reading, he contributed to reviews and protocols on ecosystem services for poverty alleviation, focusing on pollinator diversity and crop interactions in urban and agricultural landscapes.

Dr. Otieno's consultancy roles have been pivotal in developing conservation tools and conducting ecological surveys. He worked with the National Museums of Kenya to enhance sustainable agriculture through pollinator community management and with the Biodiversity Conservation Programme funded by the EU on projects impacting local livelihoods in community wildlife sanctuaries.

Internationally recognized for his contributions, Dr. Otieno has held several notable appointments. Since 2022, he has been the Director of Research and Extension at the University of Embu and serves as the External Linkages Liaison Officer, fostering collaborations with institutions worldwide. His previous roles include Dean of Students, Secretary of the Student Welfare Committee, and member of the University Senate, showcasing his dedication to academic governance and student welfare.

Dr. Otieno's research excellence is highlighted by numerous grants and awards. He secured funding from diverse sources, including the German Research Foundation (DFG), Alexander von Humboldt Foundation, JRS Biodiversity Foundation, British Ecological Society, and USDA. His projects range from enhancing pest management in post-COVID-19 farming to studying pollinator interactions in orphan legume production across various ecosystems.

His scholarly output includes over twenty refereed journal articles, reflecting the impact of his research on global agricultural practices. His work on pollination supply models, plant–pollinator interactions, and crop pest management has been published in top-tier journals like PLoS ONE, Ecography, Science, Nature Human Behaviour, and Environmental Advances. He has also contributed to pivotal conferences, sharing insights on agroecology, ecosystem services, and pollination biology, and facilitating workshops to enhance research skills among academics.

Dr. Otieno's role as an educator extends to supervising numerous postgraduate and undergraduate students. He has guided Ph.D. and MSc candidates through groundbreaking research projects, fostering the next generation of scientists and innovators. His international teaching assignments, in collaboration with institutions like ETH Zürich, University of Würzburg, and the Tropical Biology Association, further highlight his commitment to knowledge dissemination and capacity building. In addition to his academic and professional pursuits, Dr. Otieno is actively involved in editorial services and peer reviewing for several scientific journals. As an Associate Editor for the Journal of Applied Ecology and guest editor for Frontiers in Sustainable Food Systems, he supports the dissemination of research findings in his fields of expertise.

Dr. Otieno is a member of several professional societies, including the British Ecological Society, The Entomological Society of America, and The Society for Conservation Biology, reflecting his engagement with the global scientific community. His collaborative efforts span continents, with projects involving institutions in Germany, Israel, Argentina, the USA, Switzerland, Finland, and Kenya, among others.

Chapter 1
Introduction

1.1 Definition of Agroecology

Agroecology is an environmentally beneficial method of agricultural production that integrates ecological principles and knowledge with social and economic factors (Gliessman 1995). This nearly four-decade-old subject emphasizes the importance of sustainable agricultural practices in supporting nature's resilience to environmental or human-induced shocks, biodiversity, and soil health (Suárez-Torres et al. 2017). The philosophy of agroecology is achieving resilient, productive, and economically viable agricultural systems that are socially and environmentally responsible.

Indeed, agroecology is not just a collection of farming methods, but it serves as a transforming theory that seeks to integrate agriculture and nature's cycles into harmony for mutual benefits of natural ecosystems and human life (Suárez-Torres et al. 2017). The principles of agroecology stress that specific sites must be understood in order to preserve them, for instance, distinct ecosystems or societies (Gliessman 1995; Francis et al. 2003; Suárez-Torres et al. 2017; Barrios et al. 2020).

Agroecology has emerged as a key lever for transforming food and agriculture systems to address interconnected economic, environmental, and social challenges. By paying attention to how, where, and for whom food is produced, agroecology offers a path toward building more resilient and equitable food systems.

1.2 Agroecological Context in Sub-Saharan Africa

Sub-Saharan Africa is facing the detrimental impacts of climate change, including variations in temperature, rainfall patterns, and more frequent extreme weather events. To address these challenges and ensure food security, the adoption of

M. Otieno, *Sustainable Agroecological Practices in Sub-Saharan Africa in the Face of Climate Change*, Advances in Geographical and Environmental Sciences, https://doi.org/10.1007/978-3-031-70472-7_1

sustainable agroecological approaches is crucial (Jayne et al. 2018). Agroecological practices can enhance the resilience and sustainability of agricultural systems in the region. One such approach is the incorporation of indigenous knowledge and practices into farming. Research suggests that local communities in Sub-Saharan Africa possess valuable knowledge and strategies for coping with the effects of climate change on their agricultural activities (Apraku et al. 2021). This knowledge includes techniques for predicting seasonal weather patterns, managing wind speeds, and preserving crops for planting, all of which can contribute to the development of more resilient and sustainable farming systems.

Additionally, building the skills, knowledge, and capacity of rural communities is essential for modernizing agricultural systems and making them more climate-smart. Capacity-building efforts can empower farmers to adopt innovative practices and technologies that improve the productivity and nutrient content of Afrotropical crops, thereby enhancing food security, nutrition, and health in the face of climate change (Devenish et al. 2023).

Some of the sustainable agroecological approaches that are applicable to Sub-Saharan Africa in mitigating the effects of climate change include;

(i) Conservation agriculture

Conservation agriculture has gained significant attention in recent years as a sustainable farming approach in sub-Saharan Africa. This practice involves minimal soil disturbance, the use of cover crops or crop residues, and crop rotation or diversification (Nyagumbo et al. 2020; Antwerpen et al. 2021). These principles help to boost soil properties crucial for agriculture, such as moisture retention, erosion reduction, and improved soil health, thereby enhancing the resilience of farming systems to climate variability.

The benefits of conservation agriculture have been well-documented in the region. A synthesis of seven-year maize yield data from five Eastern and Southern African countries found that conservation agriculture systems outperformed conventional farming practices, with a significant reduction in yield variability (Nyagumbo et al. 2020). Additionally, case studies from South Africa have highlighted the importance of conservation agriculture in addressing the country's water scarcity and limited arable land, through measures such as ripping, mulching, and crop rotation (Swanepoel et al. 2018; Antwerpen et al. 2021). Furthermore, a study in Tanzania has demonstrated the potential of conservation agriculture technology packages to enhance the productivity, resilience, and sustainability of smallholder maize-legume cropping systems, in the face of challenges posed by unreliable rainfall, short cropping seasons, and climate variability (Nyagumbo et al. 2020; Sariah and Rodriguez 2020).

Despite these successes, the widespread adoption of conservation agriculture in sub-Saharan Africa remains a challenge. As noted in a review of soil and water conservation in the region, more work is needed to scale up local success stories and demonstrate the full contribution of these practices to poverty alleviation and food security (Zhang et al. 2021).

(ii) Agroforestry

Agroforestry has emerged as a promising approach to address the pressing challenges faced by sub-Saharan African communities, offering a multifaceted solution to environmental, agricultural, and socioeconomic concerns. By integrating trees with crops and livestock systems, agroforestry provides a wide range of benefits, including shade, windbreaks, and nutrient cycling, which contribute to soil conservation, microclimate regulation, and biodiversity conservation (Nair 2014; Wilson and Lovell 2016). Moreover, agroforestry diversifies income sources and enhances resilience to the impacts of climate change, making it an invaluable practice in the region.

The advantages of agroforestry are well-documented, with studies showing its potential to address various land-use challenges. Agroforestry practices, such as tree intercropping, home gardens, shaded perennial cash crops, silvopasture, windbreaks, and tree woodlots, have been recognized for their ability to improve food and nutritional security while mitigating environmental degradation (Nair 2007, 2014). These practices offer sustainable solutions to issues like food security, environmental protection, and climate change mitigation, both in developing countries and the industrialized world.

As the benefits of agroforestry become increasingly clear, modern land-use systems are evolving toward a more sustainable and holistic approach to land management. The choice of tree species used in agroforestry depends on various environmental, social, and economic factors, with recent development programs in the Sahel region of Africa focusing on the use of leguminous trees, such as acacia, to combat desertification.

Despite the growing recognition of agroforestry's potential, more research and effort are needed to explore its full scope and to raise awareness of its benefits. Agroforestry can offer a promising solution to the complex challenges faced by sub-Saharan African communities by harnessing the sustainability attributes and production benefits of the time-tested practices of growing trees and crops, sometimes with animals, in interacting combinations (Nair 2007, 2014; Koskey et al. 2021).

(iii) Water harvesting and storage

Water scarcity is a significant challenge for agricultural production, human consumption, and overall development in sub-Saharan Africa. Developing water harvesting techniques, such as rainwater harvesting, small-scale reservoirs, and ponds, can help capture and store water during the rainy season for use during dry periods. This approach improves water availability for crops, livestock, and household use, especially in areas with erratic rainfall patterns (Shadeed et al. 2020).

Water harvesting refers to the collection and storage of rainwater or runoff for later use. This practice is critical in sub-Saharan Africa, where absolute water scarcity and the inability of crops to use available water are major technical issues (Stroosnijder et al. 2012). Significant progress has been made in identifying land management practices that improve water-use efficiency, such as enhancing infiltration and storage of rainwater, reducing evaporation losses, and using treated wastewater for irrigation. In the West Bank, rainwater harvesting is considered a strategic alternative to satisfy the increasing water supply-demand gap for both domestic and agricultural uses (Shadeed et al. 2020).

Domestic rainwater harvesting, which provides water directly to households, enables a number of small-scale productive activities and has the potential to supply water even in rural and peri-urban areas that conventional technologies cannot reach. The South African government has committed to providing financial assistance to poor households for the capital cost of rainwater storage tanks and related works in the rural areas, as part of the effort to achieve the Millennium Development Goals. However, the legal status of domestic rainwater harvesting remains unclear, and there are risks of waterborne diseases associated with improper use of the system.

Water harvesting and storage techniques are essential in sub-Saharan Africa, where drylands show a wide diversity and, therefore, require appropriately adapted best mitigation practices and strategies (Kahinda et al. 2007; Stroosnijder et al. 2012; Shadeed et al. 2020).

(iv) Drought-tolerant crop varieties

Selecting and promoting drought-tolerant crop varieties adapted to local conditions is essential for ensuring productivity in the face of water scarcity. Breeding programs and seed systems should prioritize the development and dissemination of drought-resistant crop varieties suitable for sub-Saharan Africa. Drought-tolerant crop varieties play a crucial role in sub-Saharan Africa, where recurrent droughts are a major challenge to agricultural productivity and food security. These crop varieties are specifically bred or selected to withstand periods of low rainfall and water stress (Nepolean et al. 2018; Senyolo et al. 2018; Al-Ashkar et al. 2021). Plant breeding has proven successful in sustained improvement of drought tolerance in rainfed crops and dryland crops throughout the years, and researchers are seeking credible screening criteria for drought tolerance in wheat genotypes (Al-Ashkar et al. 2021).

Several varieties and hybrids in different food crops including maize have been released through conventional breeding, but the ever-growing population, decreasing agricultural land, lowering water table, changing climate, and other variables pose tremendous challenges to researchers to improve the production and productivity of food crops. Drought is one of the major problems to sustain and improve the productivity of food crops including maize in tropical and subtropical production systems, and with the advent of novel genomics and breeding tools, the way of doing breeding has been tremendously changed in the last two decades (Nepolean et al. 2018). Drought tolerance is a combination of several component traits with a quantitative mode of inheritance (Nepolean et al. 2018; Khadka et al. 2020; Menkir et al. 2020; Al-Ashkar et al. 2021).

Breeding new high-yielding drought-tolerant wheat varieties is a research priority specifically in regions where climate change is predicted to result in more drought conditions. A strategy that evaluates genotypes for physiological responses to drought at earlier growth stages may be more targeted to drought and time efficient. Such an approach may be enabled by recent advances in high-throughput phenotyping platforms. Commonly in breeding for drought tolerance, grain yield is the basis for selection, but it is a complex, late-stage trait, affected by many factors aside from drought.

Clusters associated with improved drought tolerance consisted of phenotypic states that likely enable greater soil exploration by reallocating internal resources to greater root construction, restrict uptake of water to conserve soil moisture, and improve penetrability of hard, dry soils. We propose that the most drought-tolerant–integrated phenotypes merit consideration as breeding ideotypes (Klein et al. 2020).

(v) Agroecological pest and disease management

Agroecological approaches to pest and disease management have gained prominence in sub-Saharan Africa as sustainable alternatives to conventional pesticide-based methods. These approaches emphasize ecological principles, biodiversity conservation, and holistic farming practices, with a focus on managing pests and diseases through natural methods such as biological control, crop diversification, and habitat manipulation (Barzman et al. 2015; Sutherland et al. 2017; Harrison et al. 2019).

The over-reliance on chemical control has led to the contamination of ecosystems and undesirable health effects, as well as the emergence of pest resistance and declining availability of active substances. Consequently, there is a growing need to design cropping systems that are less dependent on synthetic pesticides. Integrated Pest Management (IPM) principles, which fit within sustainable farm management, offer a dynamic and flexible approach that accounts for the diversity of farming situations and the complexities of agroecosystems, thereby improving the resilience of cropping systems and our capacity to adapt crop protection to local realities (Barzman et al. 2015).

Agroecological approaches, such as the use of biological control and resistant crop varieties, have been shown to be effective in reducing pest infestations and increasing food production by African farmers (Odindo 1991). Recent advances in genetic screening and engineering are also enabling the widespread use of biological pest control or growth stimulation treatments based on bacteria or fungi, which are viewed by the agrochemical industry as an alternative to synthetic compounds and an area of potential commercial growth. The deployment of these agroecological options, however, requires an understanding of current smallholder pest management strategies and the way these are constrained by farm management and livelihood factors.

Smallholder farmers in Africa and Asia already contend with a variety of crop pests, including several native armyworm species, which can reduce yields by approximately 30% across the continent (Harrison et al. 2019). The arrival of the fall armyworm (*Spodoptera frugiperda*) has further exacerbated the pest management challenges faced by these farmers, necessitating the exploration of alternative, ecologically based control methods.

Demonstration plot showing simple techniques for pest control

(v) Soil and nutrient management

Soil and nutrient management are a critical component of ensuring sustainable agricultural production and food security in sub-Saharan Africa (SSA). Many agricultural systems in the region suffer from soil degradation, nutrient deficiencies, and low fertility levels, which hinder crop growth and productivity (Mndzebele et al. 2020; Xiong and Tarnavsky 2020). Effective soil and nutrient management practices, such as organic matter addition, composting, and biological nitrogen fixation, can enhance soil health and resilience, thereby improving agricultural productivity and promoting sustainable farming systems (Tenywa and Bekunda 2009; Xiong and Tarnavsky 2020).

The hyacinth beans (*Lablab purpureus*), a legume crop rich in nitrogen

Nutrient-poor soils coupled with micronutrient deficiency remain a significant challenge for many rural, resource-poor communities in sub-Saharan Africa

(Mndzebele et al. 2020). Advances in food production are further constrained by the "invisibility factor," where women, who are the primary smallholder farmers in the region, have limited access to the necessary resources and inputs to increase yields and output (Gladwin et al. 1997). To address this issue, better collection of gender-disaggregated data and the implementation of policies and programs that consider the specific needs and constraints faced by women farmers are essential.

Several technical options for soil improvement and restoration are available, but the adoption of these practices at scale has been hindered by sociocultural, institutional, economic, and policy barriers. Gender inequality is deeply embedded in these barriers and represents a complex challenge that requires context-specific understandings and approaches. Successful local soil and water conservation activities practiced by farmers have been documented, but the challenge remains to scale these successes to wider implementation (Tenywa and Bekunda 2009).

(vi) Farmer field schools and knowledge exchange

Farmer Field Schools (FFS) and knowledge exchange initiatives have emerged as powerful platforms for driving agricultural development and enhancing the capacity of smallholder farmers in sub-Saharan Africa. These platforms facilitate the sharing of agroecological practices, traditional knowledge, and climate-smart farming techniques, empowering farmers to adapt and innovate in the face of climate change (Friis-Hansen 2008; Dolinska et al. 2020; Abdulsamad et al. 2013).

Enhancing the ability of farmers to connect with knowledge, networks, and institutions is a fundamental development challenge in most lower and middle-income countries, as farmers often struggle to ensure the socioeconomic well-being of their dependents and future generations (Abdulsamad et al. 2013). Farmer Field Schools and knowledge exchange initiatives provide a unique opportunity to address this challenge by promoting practical skills, sustainable agricultural practices, and enhanced decision-making abilities among farmers (Friis-Hansen 2008; Dolinska et al. 2020; Gangwar et al. 2020; Muriithi et al. 2021).

Research from Soroti district, Uganda, demonstrates the significant impact of a combination of farmer empowerment and innovation through experiential learning in Farmer Field School groups, changes in the opportunity structure through transformation of local government staff, establishment of new farmer-governed local institutions, and the emergence of private service providers (Friis-Hansen 2008). This approach has been successful in reducing rural poverty through the market-based spread of pro-poor agricultural technologies, highlighting the importance of an institutional setting that combines farmer empowerment with an enabling policy environment.

Similarly, the EAU4 Food project, which aimed to increase food production in irrigation schemes in Africa through improved farming strategies, utilized a platform design inspired by the community of practice concept. This approach opened up space for farmers' interactive learning and enabled their active participation in the innovation process, allowing them to contribute to the identification of innovation needs and the conduct of experiments (Friis-Hansen 2008; Dolinska et al. 2020; Gangwar et al. 2020).

As the agriculture sector in sub-Saharan Africa evolves, with private entrepreneurs coordinating extensive value chains and new providers of knowledge from the private sector and civil society, Farmer Field Schools and knowledge exchange platforms can play a crucial role in empowering smallholder farmers to adapt and innovate (Friis-Hansen 2008; Dolinska et al. 2020; Gangwar et al. 2020). These initiatives can help bridge the gap between innovative solutions and the economic marginalization of certain farmers, ensuring that the benefits of agricultural development are equitably distributed.

(vii) Climate-smart livestock management

The issue of climate change poses a significant threat to the agricultural sector, particularly the livestock industry, in sub-Saharan Africa. The region is expected to experience a warming trend, increased aridity, and changes in rainfall patterns, which can have detrimental effects on the productivity and nutrient content of crops and livestock. These environmental challenges are likely to disproportionately impact pastoral communities whose livelihoods are heavily dependent on livestock production (Machan et al. 2022).

The camel milk value chain in the arid and semi-arid regions of Africa, such as the Sahel and Horn of Africa, has the potential to provide a sustainable source of income and nutrition for pastoralist communities (Machan et al. 2022). However, the sustenance of this value chain is threatened by the effects of climate change, including abrupt changes in weather conditions, unreliable rainfall patterns, and extreme weather events like drought and flooding (Machan et al. 2022). In the Somali region of Ethiopia, for example, the large population that relies on livestock and crop production for subsistence is highly vulnerable to the impacts of climate change (Devenish et al. 2023).

Addressing the threats posed by climate change to livestock production in sub-Saharan Africa will require a multifaceted approach that focuses on building the skills, knowledge, and capacity of rural communities to modernize and adapt their agricultural systems (Girmay et al. 2018; Oduniyi et al. 2020; Machan et al. 2022; Devenish et al. 2023). Strategies may include the promotion of climate-smart livestock management practices such as improved feed and water management, disease prevention, and the diversification of livestock species. Additionally, investment in infrastructure, extension services, and early warning systems can help to build the resilience of pastoral communities to the effects of climate change (Girmay et al. 2018; Oduniyi et al. 2020; Machan et al. 2022; Devenish et al. 2023).

(viii) Integrated water management

Efficient and integrated water management practices have become increasingly crucial in addressing the complex water-related challenges faced by communities in sub-Saharan Africa. These practices, including precision irrigation, water recycling, and efficient water-use techniques, can help conserve water resources and reduce water-related risks in the region. Integrated water management approaches consider the entire water cycle, from source to consumption, and promote sustainable water use (Tenywa and Bekunda 2009; Larsen et al. 2016).

Sustainable water management in sub-Saharan Africa involves a holistic approach to managing water resources and water-related issues in a coordinated and sustainable manner. This includes considering the interconnectedness of water availability, water quality, water use, and water governance to ensure the efficient and equitable management of water resources (Tenywa and Bekunda 2009; Wang et al. 2014). Sub-Saharan Africa continues to face significant challenges, such as destructive extraction, over-exploitation, and inadequate conservation of soil and water resources. Efforts are needed to scale up successful local soil and water conservation activities practiced by farmers in the region.

Emerging solutions, such as new concepts for stormwater drainage, increased water productivity, distributed or on-site treatment of wastewater, and source separation of human waste, have the potential to provide locally adapted, resource-efficient alternative solutions for urban water sustainability in the rapidly growing urban areas of sub-Saharan Africa (Larsen et al. 2016). These innovative approaches, combined with institutional and organizational reforms, can help address the pressing water challenges in the region and promote long-term water security.

(ix) Farmer empowerment and policy support

Supporting smallholder farmers through policy interventions, access to credit, extension services, and market opportunities is crucial for promoting sustainable agroecological practices. Government support and investment in research, infrastructure, and capacity building can help create an enabling environment for farmers to adopt climate-resilient and sustainable agricultural practices. Farmer empowerment and policy support are essential elements for agricultural development and food security in sub-Saharan Africa, as they play a significant role in enabling smallholder farmers to overcome challenges, improve their livelihoods, and contribute to sustainable agricultural practices.

The South African government has been assisting small-scale farmers financially to alleviate poverty, create jobs, and improve food security by promoting small-scale commercial farming, primarily in rural areas, through initiatives such as the Micro-Agriculture Finance Institutions of South Africa and the Department of Agriculture, Forestry, and Fisheries (Muzekenyi et al. 2023). However, there is no clear strategy for achieving significant and evidence-based development (Hlatshwayo et al. 2021; Muzekenyi et al. 2023). Public support programs have been developed as part of the Black Economic Empowerment strategy to improve the performance of emerging farmers, which requires upgrading their skills, providing access to suitable agricultural land and water, and ensuring the presence of various factors such as marketing and service institutions, input markets, and accessible output markets (Schalkwyk et al. 2011; Muzekenyi et al. 2023). Channeling support to this sector would have a positive, albeit limited, impact on rural poverty, mainly through raising employment, with a smaller impact on wages due to the slackness in the labor market.

If state support could be effectively channeled to supporting viable land reform and small-scale agriculture, this is likely to be much more effective as a pro-poor strategy compared to supporting urban and industrial development, which will tend to benefit enterprises and a smaller class of better-paid workers first (Schalkwyk et al. 2011; Black and Gerwel 2014; Muzekenyi et al. 2023). Understanding the small-scale agricultural sector as a precondition for promoting rural development is crucial, as this form of farming is dominant in the country's rural areas where at least 70% of the country's poorest households' dwell, and can stimulate rural development as it is labor-intensive, translating to high employment opportunities (Schalkwyk et al. 2011; Black and Gerwel 2014; Tshuma 2014; Muzekenyi et al. 2023). Therefore, a clear strategy for promoting small-scale commercial farming as a strategy for rural economic development in South Africa is needed, including defining the development strategy, delineating production, performance, policy, and control measures (Tshuma 2014; Muzekenyi et al. 2023).

References

Abdulsamad A, Brun LC, Gereffi G (2013) Realizing the potential of African agriculture: innovations and market access for smallholder farmers. Duke University, Center on Globalization, Governance & Competitiveness, May 2013, https://doi.org/10.2139/ssrn.2598875

Al-Ashkar I, Al-Suhaibani N, Abdella KA, Sallam M, Alotaibi MA, Seleiman MF (2021) Combining genetic and multidimensional analyses to identify interpretive traits related to water shortage tolerance as an indirect selection tool for detecting genotypes of drought tolerance in wheat breeding. Plan Theory 10(5):931–931. https://doi.org/10.3390/plants10050931

Antwerpen RV, Laker MC, Beukes DJ, Botha JJ, Collett A, Plessis MD (2021) Conservation Agriculture farming systems in rainfed annual crop production in South Africa. S Afr J Plant Soil 38(3):202–216. https://doi.org/10.1080/02571862.2020.1797195

Apraku A, Morton J, Gyampoh BA (2021) Climate change and small-scale agriculture in Africa: does indigenous knowledge matter? Insights from Kenya and South Africa. Sci African 12:e00821. https://doi.org/10.1016/j.sciaf.2021.e00821

Barrios E, Gemmill-Herren B, Bicksler A, Siliprandi E, Brathwaite R, Moller S et al (2020) The 10 elements of agroecology: enabling transitions towards sustainable agriculture and food systems through visual narratives. Ecosyst People 16(1):230–247. https://doi.org/10.1080/26395916.2020.1808705

Barzman M, Bàrberi P, Birch ANE, Boonekamp P, Dachbrodt-Saaydeh S, Graf B et al (2015) Eight principles of integrated pest management. Agron Sustain Dev 35:1199–1215. https://doi.org/10.1007/s13593-015-0327-9

Black A, Gerwel H (2014) Shifting the growth path to achieve employment intensive growth in South Africa. Dev South Afr 31(2):241–256. https://doi.org/10.1080/0376835x.2013.871198

Devenish AJ, Schmitter P, Jellason NP, Esmail N, Abdi NM, Adanu SK et al (2023) One hundred priority questions for the development of sustainable food systems in sub-Saharan Africa. Land 12(10):1879. https://doi.org/10.22541/au.169419355.58766387/v1

Dolinska A, Oates N, Ludi E, Habtu S, Rougier JE, Sanchez-Reparaz M et al (2020) Engaging farmers in a research project. Lessons learned from implementing the Community of Practice Concept in innovation platforms in irrigated schemes in Tunisia. Mozambique and Ethiopia Irrigation and Drainage 69:38–48. https://doi.org/10.1002/ird.2222

Francis C, Lieblein G, Gliessman S, Breland TA, Creamer N, Harwood R et al (2003) Agroecology: the ecology of food systems. J Sustain Agric 22(3):99–118. https://doi.org/10.1300/j064v22n03_10

Friis-Hansen E (2008) Impact assessment of farmer institutional development and agricultural change: Soroti district. Uganda Participat Res Gender Anal 18(4–5):506–523. https://doi.org/10.1080/09614520802181236

Gangwar DS, Tyagi S, Soni SK (2020, February) Connecting farmers to knowledge, networks and institutions for agroecological sustainability. In: 2020 International Conference on Electrical and Electronics Engineering (ICE3). IEEE, pp 311–315. https://doi.org/10.1109/ice348803.2020.9122983

Girmay FD, Gebreselassie G, Bajigo A (2018) Climate change risk management and coping strategies for sustainable camel production in the case of Somali Region. Ethiopia J Biotechnol Res 4(9):66–75. https://doi.org/10.32861/jbr.49.66.75

Gladwin CH, Buhr KL, Goldman A, Hiebsch C, Hildebrand PE, Kidder G et al (1997) Gender and soil fertility in Africa. Replenishing Soil Fertil Africa 51:219–236. https://doi.org/10.2136/sssaspecpub51.c9

Gliessman SR (1995) 3 Sustainable agriculture: an agroecological perspective. In: Advances in plant pathology, vol 11. Academic Press, pp 45–57. https://doi.org/10.1016/s0736-4539(06)80005-x

Harrison RD, Thierfelder C, Baudron F, Chinwada P, Midega C, Schaffner U, Van Den Berg J (2019) Agro-ecological options for fall armyworm (*Spodoptera frugiperda* JE Smith) management: providing low-cost, smallholder friendly solutions to an invasive pest. J Environ Manag 243:318–330. https://doi.org/10.1016/j.jenvman.2019.05.011

Hlatshwayo SI, Ngidi M, Ojo T, Modi AT, Mabhaudhi T, Slotow R (2021) A typology of the level of market participation among smallholder farmers in South Africa: Limpopo and Mpumalanga Provinces. Sustain For 13(14):7699. https://doi.org/10.3390/su13147699

Jayne TS, Sitko NJ, Mason NM, Skole D (2018) Input subsidy programs and climate smart agriculture: current realities and future potential. Clim Smart Agric 251–273. https://doi.org/10.1007/978-3-319-61194-5_12

Kahinda JMM, Taigbenu AE, Boroto JR (2007) Domestic rainwater harvesting to improve water supply in rural South Africa. Phys Chem Earth Parts a/b/c 32(15–18):1050–1057. https://doi.org/10.1016/j.pce.2007.07.007

Khadka K, Earl HJ, Raizada MN, Navabi A (2020) A physio-morphological trait-based approach for breeding drought tolerant wheat. Front Plant Sci 11:511313. https://doi.org/10.3389/fpls.2020.00715

Klein SP, Schneider HM, Perkins AC, Brown KM, Lynch JP (2020) Multiple integrated root phenotypes are associated with improved drought tolerance. Plant Physiol 183(3):1011–1025. https://doi.org/10.1104/pp.20.00211

Koskey G, Mburu SW, Awino R, Njeru EM, Maingi JM (2021) Potential use of beneficial microorganisms for soil amelioration, phytopathogen biocontrol, and sustainable crop production in smallholder agroecosystems. Front Sustain Food Syst 5:606308. https://doi.org/10.3389/fsufs.2021.606308

Larsen TA, Hoffmann S, Lüthi C, Truffer B, Maurer M (2016) Emerging solutions to the water challenges of an urbanizing world. Science 352(6288):928–933. https://doi.org/10.1126/science.aad8641

Machan SN, Agwata JF, Oguge NO (2022) Environmental factors influencing the sustenance of the camel milk value chain in Isiolo County. Northern Kenya Resourc 11(3):27. https://doi.org/10.3390/resources11030027

Menkir A, Crossa J, Meseka S, Bossey B, Muhyideen O, Riberio PF et al (2020) Stacking tolerance to drought and resistance to a parasitic weed in tropical hybrid maize for enhancing resilience to stress combinations. Front Plant Sci 11:166. https://doi.org/10.3389/fpls.2020.00166

Mndzebele B, Ncube B, Nyathi M, Kanu SA, Fessehazion M, Mabhaudhi T et al (2020) Nitrogen fixation and nutritional yield of cowpea-amaranth intercrop. Agronomy 10(4):565. https://doi.org/10.3390/agronomy10040565

Muriithi LN, Onyari CN, Mogaka HR, Gichimu BM, Gatumo GN, Kwena K (2021) Adoption determinants of adapted climate smart agriculture technologies among smallholder farmers

in Machakos, Makueni, and Kitui Counties of Kenya. J Agric Exten 25(2):75–85. https://doi.org/10.4314/jae.v25i2.7
Muzekenyi M, Nyika F, Hoque M (2023) A small-scale farming intervention plan for inclusive economic development in rural South Africa. Int J Food Agric Nat Resourc 4(2):46–52. https://doi.org/10.46676/ij-fanres.v4i2.136
Nair PR (2007) The coming of age of agroforestry. J Sci Food Agric 87(9):1613–1619. https://doi.org/10.1002/jsfa.2897
Nair PKR (2014) Agroforestry: practices and systems, an introduction to agroforestry, 270–282. https://doi.org/10.1016/b978-0-444-52512-3.00021-8
Nepolean T, Kaul J, Mukri G, Mittal S (2018) Genomics-enabled next-generation breeding approaches for developing system-specific drought tolerant hybrids in maize. Front Plant Sci 9:325697. https://doi.org/10.3389/fpls.2018.00361
Nyagumbo I, Mupangwa W, Chipindu L, Rusinamhodzi L, Craufurd P (2020) A regional synthesis of seven-year maize yield responses to conservation agriculture technologies in Eastern and Southern Africa. Agric Ecosyst Environ 295:106898–106898. https://doi.org/10.1016/j.agee.2020.106898
Odindo MO (1991) Pest and vector management in Africa International Journal of Tropical Insect. Science 12(5–6):527–533. https://doi.org/10.1017/s1742758400012984
Oduniyi OS, Rubhara TT, Antwi MA (2020) Sustainability of livestock farming in South Africa. Outlook on production constraints, climate-related events, and upshot on adaptive capacity. Sustain For 12(7):2582. https://doi.org/10.3390/su12072582
Sariah J, Rodriguez D (2020, April) Technology package for sustainable productivity and resiliency to climate variability: the proof from SIMLESA studies in Tanzania. In: Proceedings (Vol 36, No 1, p 168). MDPI. https://doi.org/10.3390/proceedings2019036168
Schalkwyk HDV, Obi A, Tilburg AV, Fraser G (2011) Unlocking markets to smallholders: lessons from South Africa. Wageningen Academic Publishers, p 268. https://library.oapen.org/bitstream/20.500.12657/34544/1/413359.pdf
Senyolo MP, Long TB, Blok V, Omta O (2018) How the characteristics of innovations impact their adoption: an exploration of climate-smart agricultural innovations in South Africa. J Clean Prod 172:3825–3840. https://doi.org/10.1016/j.jclepro.2017.06.019
Shadeed S, Judeh T, Riksen M (2020) Rainwater harvesting for sustainable agriculture in high water-poor areas in the West Bank. Palestine Water 12(2):380. https://doi.org/10.3390/w12020380
Stroosnijder L, Moore D, Alharbi A, Argaman E, Biazin B, van den Elsen E (2012) Improving water use efficiency in drylands. Curr Opin Environ Sustain 4(5):497–506. https://doi.org/10.1016/j.cosust.2012.08.011
Suárez-Torres J, Suárez-López JR, López-Paredes D, Morocho H, Cachiguango-Cachiguango LE, Dellai W (2017) Agroecology and health: Lessons from indigenous populations. Curr Environ Health Reports 4:244–251. https://doi.org/10.1007/s40572-017-0146-z
Sutherland WJ, Barnard P, Broad S, Clout M, Connor B, Côté IM et al (2017) A 2017 horizon scan of emerging issues for global conservation and biological diversity. Trends Ecol Evol 32(1):31–40. https://doi.org/10.1016/j.tree.2016.11.005
Swanepoel CM, Swanepoel LH, Smith HJ (2018) A review of conservation agriculture research in South Africa. S Afr J Plant Soil 35(4):297–306. https://doi.org/10.1080/02571862.2017.1390615
Tenywa MM, Bekunda M (2009) Managing soils in Sub-Saharan Africa: challenges and opportunities for soil and water conservation. J Soil Water Conserv 64(1):44A–48A. https://doi.org/10.2489/jswc.64.1.44a
Tshuma MC (2014) Understanding the small-scale agricultural sector as a precondition for promoting rural development in South Africa. Afr J Agric Res 9(31):2409–2418
Wang H, Wang T, Zhang B, Li F, Toure B, Omosa IB et al (2014) Water and wastewater treatment in Africa–current practices and challenges. Clean Soil Air Water 42(8):1029–1035. https://doi.org/10.1002/clen.201300208

Wilson MH, Lovell ST (2016) Agroforestry—the next step in sustainable and resilient agriculture. Sustain For 8(6):574. https://doi.org/10.3390/su8060574

Xiong W, Tarnavsky E (2020) Better agronomic management increases climate resilience of maize to drought in Tanzania. Atmos 11(9):982. https://doi.org/10.3390/atmos11090982

Zhang W, Elias M, Meinzen-Dick R, Swallow K, Calvo-Hernandez C, Nkonya E (2021) Soil health and gender: why and how to identify the linkages. Int J Agric Sustain 19(3–4):269–287. https://doi.org/10.1080/14735903.2021.1906575

Chapter 2
Water-Use Efficiency, Water Quality, and Irrigation

2.1 Water-Use Efficiency in Agroecological Systems

Agroecology emphasizes high water-use efficiency, with maximization of agricultural productivity while minimizing water use or optimizing the same. Here it is recognized that water is limited and precious; hence, increasing water-use efficiency may decrease water loss, increase crop production per irrigated unit area, and reduce the environmental degradation arising from irrigation.

Improvement in soil health and structure is very essential for enhancing water-use efficiency. The agroecological practice of adding organic matter, cover cropping, and reduced tillage aids in keeping moisture within the soil and generally enhances the ability of the soil to hold water. This is because organic matter acts as a sponge, holding onto moisture to reduce eventually the loss of water through drainage (Yan et al. 2022). Crop selection and diversification can lead to water-use efficiency by choosing crops better corresponding to the locality's climatic conditions and availability of water (Pérez-Blanco et al. 2020). That is, in respect to less well-adapted varieties, these plants will have better uses of water (He et al. 2018). In addition, it is possible to achieve optimal water use by crop diversification because diversification decreases the risk of water stress and enables exploitation of rooting depths and water uptake patterns that differ between crops.

M. Otieno, *Sustainable Agroecological Practices in Sub-Saharan Africa in the Face of Climate Change*, Advances in Geographical and Environmental Sciences, https://doi.org/10.1007/978-3-031-70472-7_2

Hydroponic system: Growing crops under significantly less water compared to traditional soil-based agriculture because, in this system, water can be recirculated and reused

Applying organic or synthetic mulch on the soil surface is another effective strategy for improving water-use efficiency (Stroosnijder et al. 2012). Mulch helps reduce evaporation, increase infiltration, and maintain soil moisture (Stroosnijder et al. 2012). In drylands, where water scarcity is a significant challenge, adopting practices that improve infiltration and storage of rainwater, reduce evaporation losses, and utilize treated wastewater for irrigation can greatly enhance water-use efficiency (Stroosnijder et al. 2012; Yan et al. 2022).

2.2 Water Quality Management in Agroecology

Water quality is a critical factor in agroecology, as it directly impacts the health of ecosystems, agricultural productivity, and human well-being (Khan and Hanjra 2008). Agroecological practices aim to promote sustainable water management and protect water quality by minimizing pollution, preserving natural water sources, and enhancing ecosystem functions.

One key aspect of water quality in agroecology is nutrient management. Agroecology emphasizes balanced nutrient management to prevent nutrient runoff and minimize water pollution (Divya and Belagali 2012; Mukiza et al. 2021). This involves optimizing fertilizer application rates, timing, and methods to match crop requirements, reducing the risk of excess nutrients reaching water bodies and causing eutrophication. Agroecological practices such as cover cropping, crop rotation, and organic matter addition help improve nutrient cycling and reduce nutrient losses (Divya and Belagali 2012; Seufert and Ramankutty 2017; Mukiza et al. 2021).

River Athi in Kenya chocked with water hyacinth (*Eichhornia crassipes*)

Moreover, agroecology promotes organic farming practices that minimize or eliminate the use of synthetic pesticides and chemical fertilizers, which helps prevent water contamination and maintain water quality. Natural pest control methods, crop diversification, and integrated pest management are employed to manage pests and diseases in an environmentally friendly manner (Seufert and Ramankutty 2017).

Establishing riparian buffer zones along streams, rivers, and water bodies is another common practice in agroecology. These vegetated strips of land act as natural filters, trapping sediments, nutrients, and pollutants before they enter waterways. Buffer zones reduce erosion, improve water infiltration, and enhance water quality by preventing contaminants from reaching aquatic ecosystems (Dale and Polasky 2007; Thorburn et al. 2013).

Soil erosion control is also a priority in agroecological systems. Agroecological practices such as cover cropping, reduced tillage, and contour farming help stabilize soil and prevent erosion, which in turn helps maintain water quality by reducing sediment and nutrient loads in water bodies (Dale and Polasky 2007; Divya and Belagali 2012; Thorburn et al. 2013; Seufert and Ramankutty 2017).

2.3 Sustainable Irrigation Practices in Agroecology

Agroecological approaches to irrigation prioritize methods that minimize water waste, enhance soil health, and optimize plant water availability. Proper irrigation scheduling based on crop water requirements and soil moisture conditions is vital in agroecology. Farmers monitor soil moisture levels using tools like moisture sensors or by assessing soil texture and appearance to enable precise timing and application of irrigation. Smart usage of irrigation systems optimizes land area productivity and conserves natural resources by reducing overall involved emission and losses in conversion components (Jain et al. 2017).

Water-efficient techniques, such as drip irrigation, are commonly employed in agroecological systems as they deliver water directly to the plant roots, minimizing evaporation and reducing water loss through surface runoff (Jabro et al. 2020; Chowdhury et al. 2023). Deficit irrigation, a water-saving strategy under which crops are exposed to a certain level of water stress, has been shown to increase crop water-use efficiency by reducing the amount of water applied without significant yield reduction.

Agroecology also emphasizes maintaining healthy soil that can retain moisture effectively. Practices such as organic matter addition, cover cropping, and reduced tillage contribute to improved soil structure, water infiltration, and moisture-holding capacity. Healthy soils with good water-holding capacity can reduce the frequency and amount of irrigation required (El-Shafei and Mattar 2022; Patanè et al. 2011; Slamini et al. 2022). Mulching the soil surface with organic materials like straw, crop residue, or compost further helps conserve soil moisture (Patanè et al. 2011; Hu 2020).

Comprehensive technical strategies that combine increased agricultural productivity and resource conservation, such as conservation tillage, mulch cultivation, limited irrigation, introduction of drought-tolerant varieties, and rotation systems, can be adjusted to local agricultural practices to promote sustainable irrigation in agroecological systems (Paz et al. 2000; Kinzli 2012; He et al. 2018).

References

Chowdhury MR, Sourav MSU, Sulaiman RB (2023) The role of digital agriculture in transforming rural areas into smart villages. In: Technology and talent strategies for sustainable smart cities: digital futures. Emerald, pp 57–78. https://doi.org/10.1108/978-1-83753-022-920231003

Dale VH, Polasky S (2007) Measures of the effects of agricultural practices on ecosystem services. Ecol Econ 64(2):286–296. https://doi.org/10.1016/j.ecolecon.2007.05.009

Divya J, Belagali SL (2012) Impact of chemical fertilizers on water quality in selected agricultural areas of Mysore district, Karnataka, India. Int J Environ Sci 2(3):1449–1458. https://doi.org/10.6088/ijes.00202030030

El-Shafei AA, Mattar MA (2022) Irrigation scheduling and production of wheat with different water quantities in surface and drip irrigation: field experiments and modelling using CROPWAT and SALTMED. Agronomy 12(7):1488. https://doi.org/10.3390/agronomy12071488

He L, Wang S, Peng C, Tan Q (2018) Optimization of water consumption distribution based on crop suitability in the middle reaches of Heihe River. Sustain For 10(7):2119. https://doi.org/10.3390/su10072119

Hu R (2020) Overview of water-saving irrigation methods in arid/semi-arid areas. In: E3S Web of conferences, vol 191. EDP Sciences, p 02001. https://doi.org/10.1051/e3sconf/202019102001

Jabro JD, Stevens WB, Iversen WM, Allen BL, Sainju UM (2020) Irrigation scheduling based on wireless sensors output and soil-water characteristic curve in two soils. Sensors 20(5):1336. https://doi.org/10.3390/s20051336

Jain P, Kumar P, Palwalia DK (2017, April) Irrigation management system with micro-controller application. In: 2017 1st International Conference on Electronics, Materials Engineering and Nano-Technology (IEMENTech). IEEE, pp 1–6. https://doi.org/10.1109/iementech.2017.8076969

Khan S, Hanjra MA (2008) Sustainable land and water management policies and practices: a pathway to environmental sustainability in large irrigation systems. Land Degrad Dev 19(5):469–487. https://doi.org/10.1002/ldr.852

Kinzli KD (2012) A low cost remote monitoring method for determining farmer irrigation practices and water use. Problems, perspectives and challenges of agricultural water management. https://doi.org/10.5772/29770

Mukiza P, Bazimenyera JDD, Nkundabose JP, Niyonkuru R, Bapfakurera NE (2021) Assessment of irrigation water quality parameters of Nyandungu Wetlands. J Geosci Environ Protect 9(10):151–160. https://doi.org/10.4236/gep.2021.910011

Patanè C, Tringali S, Sortino O (2011) Effects of deficit irrigation on biomass, yield, water productivity and fruit quality of processing tomato under semi-arid Mediterranean climate conditions. Sci Hortic 129(4):590–596. https://doi.org/10.1016/j.scienta.2011.04.030

Paz VPDS, Teodoro REF, Mendonça FC (2000) Recursos hídricos, agricultura irrigada e meio ambiente. Revista Brasileira de Engenharia Agrícola e Ambiental 4:465–473. https://doi.org/10.1590/s1415-43662000000300025

Pérez-Blanco CD, Essenfelder AH, Perry C (2020, June 1) Irrigation technology and water conservation. Rev Environ Econ Policy 14(2):216–239. https://doi.org/10.1093/reep/reaa004

Seufert V, Ramankutty N (2017) Many shades of gray—the context-dependent performance of organic agriculture. Sci Adv 3(3):e1602638. https://doi.org/10.1126/sciadv.1602638

Slamini M, Sbaa M, Arabi M, Darmous A, Bouknana D (2022) Citrus yield response and irrigation water use efficiency under partial root drying irrigation in a pilot exploitation in the Triffa plain (Eastern Morocco). Ecol Eng Environ Technol 23. https://doi.org/10.12912/27197050/143385

Stroosnijder L, Moore D, Alharbi A, Argaman E, Biazin B, van den Elsen E (2012) Improving water use efficiency in drylands. Curr Opin Environ Sustain 4(5):497–506. https://doi.org/10.1016/j.cosust.2012.08.011

Thorburn PJ, Wilkinson SN, Silburn DM (2013) Water quality in agricultural lands draining to the Great Barrier Reef: a review of causes, management and priorities. Agric Ecosyst Environ 180:4–20. https://doi.org/10.1016/j.agee.2013.07.006

Yan M, Zhang L, Ren Y, Zhang T, Zhang S, Li H et al (2022) The higher water absorption capacity of small root system improved the yield and water use efficiency of maize. Plan Theory 11(17):2300. https://doi.org/10.3390/plants11172300

Chapter 3
Current Climate, Soil, and Natural Vegetation

3.1 Current Climate

Agroecological practices have gained growing recognition as a promising approach to address the multifaceted challenges posed by climate change, food security, and environmental sustainability. Climate-smart agriculture (CSA) has emerged as a leading strategy within this broader agroecological framework, aiming to simultaneously increase productivity, enhance climate resilience, and reduce greenhouse gas emissions.

3.1.1 Climate-Smart Agriculture

At the core of climate-smart agriculture is the integration of various practices and technologies that can help farmers adapt to and mitigate the impacts of climate change. These include techniques such as diversified cropping systems, improved soil management, water-efficient irrigation, and the adoption of climate-resilient crop varieties (Mubiru et al. 2018). By fostering biodiversity, preserving water resources, and promoting sustainable land management, climate-smart agriculture seeks to strengthen the overall resilience of agricultural systems while also contributing to climate change mitigation (Rashid et al. 2021). However, the complex and interdependent nature of these objectives can lead to necessary trade-offs, as progress in one area may negatively impact another (Hellin and Fisher 2019). A key challenge in implementing climate-smart agriculture lies in the coordination of policies and programs that can effectively navigate these trade-offs and prioritize the most suitable practices for specific contexts.

A policy providing an enabling environment plays a crucial role in facilitating the adoption and scaling of climate-smart agriculture (Fusco et al. 2020). In

M. Otieno, *Sustainable Agroecological Practices in Sub-Saharan Africa in the Face of Climate Change*, Advances in Geographical and Environmental Sciences, https://doi.org/10.1007/978-3-031-70472-7_3

California, for example, a comprehensive set of policies and programs has been instrumental in driving the implementation of climate-smart practices. These efforts have focused on aligning incentives, providing technical and financial support, and fostering cross-sectoral collaboration to address the complex trade-offs inherent in climate-smart agriculture (Climate Smart Agriculture Sourcebook 2023). For instance, California's Healthy Soils Program provides grants to farmers and ranchers to implement practices that sequester carbon, reduce greenhouse gas emissions, and improve soil health. The state has also invested in research and development to identify locally adapted climate-smart solutions that can be tailored to the diverse agricultural landscapes within the state (Sova et al. 2018; Hellin and Fisher 2019; Lewis and Rudnick 2019; Rashid et al. 2021).

Similar initiatives can be found globally, where governments, development agencies, and research institutions have worked to create an enabling environment for the widespread adoption of climate-smart agriculture. In India, the National Mission for Sustainable Agriculture has been instrumental in promoting climate-resilient farming practices, such as the System of Rice Intensification and the use of drought-tolerant crop varieties.

However, realizing the full potential of climate-smart agriculture requires not only supportive policies but also the active engagement and empowerment of farmers, particularly those in vulnerable and marginalized communities (Hellin and Fisher 2019). Effective outreach, capacity building, and equitable access to resources are crucial to ensure that the benefits of climate-smart agriculture are shared across diverse agricultural systems and socioeconomic groups (Sova et al. 2018; Hellin and Fisher 2019).

3.1.2 Agroforestry

Agroforestry is a sustainable land-use practice that seamlessly integrates the cultivation of trees alongside crops or livestock within the same agricultural system. This innovative approach combines the unique benefits of trees with traditional farming activities, offering a multitude of advantages for both the environment and agricultural productivity (Nair 2007, 2013, 2014).

One of the key roles of trees in agroforestry systems is their ability to capture carbon dioxide from the atmosphere through the process of photosynthesis (Nair 2007, 2013). By sequestering carbon, trees contribute to mitigating climate change by reducing the concentration of greenhouse gases in the atmosphere. This carbon sequestration capability of trees is a crucial component of agroforestry systems, as it helps offset the carbon emissions associated with agricultural activities (Nair 2013). Furthermore, the diverse habitats created by trees in agroforestry systems serve as shelters for beneficial insects, birds, and other wildlife, which contribute to ecosystem balance by providing natural pest control, pollination services, and other ecosystem services that benefit agricultural production (Nair 2014; Abbas et al. 2017).

In addition to their environmental benefits, trees in agroforestry systems provide practical advantages for agricultural operations. They offer shade to crops and livestock, protecting them from excessive heat stress and promoting more favorable microclimatic conditions. Trees also act as natural windbreaks, reducing wind erosion and protecting crops from strong winds, thereby maintaining soil stability and preventing damage to the agricultural fields (Nair 2014, 2013, 2007). As noted in Nair (2013), such "integration and utilization of trees and shrubs that provide multiple products and services in land-use systems offer sustainable solutions to several serious land management and environmental issues such as food security, environmental protection, and climate change mitigation, in both developing countries and the industrialized world."

Furthermore, the presence of trees in agroforestry systems plays a crucial role in enhancing soil fertility. Their root systems help improve soil structure, reduce erosion, and increase organic matter content, leading to healthier and more productive soils. As trees shed leaves and organic matter, they contribute to nutrient cycling in the ecosystem, further enriching the soil for the benefit of the associated crops or livestock (Pinho et al. 2012). This synergistic relationship between the trees and the agricultural components is a hallmark of agroforestry, as it allows for the optimization of resource use and the creation of a more sustainable and resilient agricultural landscape.

The positive impacts of agroforestry extend beyond just environmental considerations. The combination of trees and crops in agroforestry systems has been shown to improve food and nutritional security, offering a sustainable alternative to monoculture production. By providing supportive and complementary roles, agroforestry can offer specific social and economic benefits across a range of landscapes and economies as well, making it a holistic approach to land management that benefits both producers and the environment (Nair 2007, 2014; Pinho et al. 2012).

An agroforestry farm with trees and crops grown together

3.1.3 Water Management

Water management plays a pivotal role in advancing climate-smart agricultural practices, as the adoption of effective water management strategies can enable farmers to optimize water utilization, enhance crop productivity, and build resilience in the face of changing climate conditions. Rainwater harvesting, drip irrigation, and precision irrigation are key components of sustainable water management in agriculture that can contribute to water conservation and sustainable farming practices.

Rainwater harvesting, which involves collecting and storing rainwater for agricultural use, can help mitigate water scarcity issues by supplementing irrigation needs during dry spells and reducing reliance on freshwater sources (He et al. 2018). This practice enhances water availability for crops, thereby contributing to sustainable farming practices. Drip irrigation, a water-efficient technique that delivers water directly to the root zone of plants in controlled amounts, can significantly improve water-use efficiency compared to traditional irrigation methods. By minimizing water wastage through evaporation and runoff, drip irrigation ensures that plants receive adequate moisture for optimal growth, contributing to water conservation and sustainable agricultural practices.

Precision irrigation, which utilizes technology and data to apply water precisely where and when it is needed in the field, can further optimize water-use efficiency. By monitoring factors such as soil moisture levels, weather conditions, and crop requirements, farmers can tailor their irrigation practices to adapt to changing rainfall patterns and maximize crop yields while conserving water resources (Bakker et al. 2008).

These water management techniques not only help improve water-use efficiency but also enable farmers to adapt to the challenges posed by variable climate conditions, such as water scarcity and unpredictable precipitation patterns. In semi-arid regions, where freshwater supplies are limited, the adoption of these strategies can be particularly beneficial in sustaining agricultural productivity and environmental quality. Optimization models and other precision agricultural technologies can be effective tools in guiding the implementation of these water management practices and achieving the water-saving potential from cultivation. By utilizing data-driven approaches to optimize planting patterns and water-land combinations, farmers can further enhance the efficiency of their water usage and adapt to changing climatic conditions. Optimization models, as highlighted in recent studies, can serve as effective tools in guiding the implementation of water management practices and identifying water-saving potentials (He et al. 2018).

Greenhouse cultivation, another important agricultural sector, can also benefit from improved water management strategies. Novel high-tech greenhouse solutions, such as closed greenhouse systems, can lead the way toward highly efficient water-use production techniques, as they can minimize water waste through reduced evaporation and enable the reuse of water resources (Kooten et al. 2008).

Healthy pepper (*Capsicum frutescens*) under greenhouse cultivation

By integrating a suite of water management strategies, including rainwater harvesting, drip irrigation, and precision irrigation, along with the adoption of advanced greenhouse technologies, farmers can optimize water utilization, enhance crop productivity, and build resilience in the face of climate change (Kooten et al. 2008; O'Connor and Mehta 2016; He et al. 2018; Ashok et al. 2020). These holistic approaches to water management, tailored to local agricultural practices and environmental conditions, can contribute significantly to the sustainability of food production systems.

For instance, studies have shown that greenhouse cultivation can reduce water demand and increase water-use efficiency compared to open-field farming (O'Connor and Mehta 2016; Ashok et al. 2020). The ability to precisely control and monitor water inputs in greenhouse systems, combined with the reduced evaporative losses, can lead to substantial water savings. Techniques such as drip irrigation, mulching, and automated irrigation scheduling in greenhouse environments have been found to improve water productivity by up to 90% relative to traditional methods (Kooten et al. 2008; O'Connor and Mehta 2016; Ashok et al. 2020). Furthermore, the integration of renewable energy systems, such as solar-powered pumps, in greenhouse operations can further enhance the sustainability of water management by reducing the reliance on energy-intensive pumping from groundwater or surface water sources.

The optimization of water-land combinations and planting patterns, as explored in various studies, can also contribute to improved water management in agricultural systems. By aligning crop selection and cultivation practices with the availability of water resources, farmers can maximize the efficient use of water and enhance the resilience of their farming systems. Ultimately, the adoption of a diverse set of water management strategies, enabled by precision agricultural technologies and tailored to local environmental and climatic conditions, can play a crucial role in advancing climate-smart and sustainable agriculture (Kooten et al. 2008; O'Connor and Mehta 2016; He et al. 2018; Ashok et al. 2020).

3.1.4 Climate-Resilient Crop Varieties

In the face of the growing challenges posed by climate change, the development and deployment of crop varieties that exhibit resilience to environmental stressors like drought, heat, and pests have become paramount for ensuring food security and sustainable agricultural production (Gupta 2010; López-Marqués et al. 2020). These climate-resilient crop varieties have the potential to serve as a crucial resource in adapting to the adverse impacts of a changing climate.

The incorporation of climate resilience traits into crop varieties can play a pivotal role in safeguarding food security and agricultural sustainability in the face of escalating environmental pressures. The continued development and widespread adoption of these climate-resilient crops are essential to equipping farmers with the tools necessary to overcome the challenges posed by the changing climate (Yumurtaç 2015; Mir et al. 2022). Genomics and reverse genetics have emerged as powerful approaches to breed resilient and nutrient-dense crops that can thrive under unfavorable climatic conditions (Mir et al. 2022). Through the strategic deployment of these advanced breeding techniques, researchers can identify and introduce beneficial traits that enhance a crop's ability to withstand environmental stressors, ensuring reliable yields even in the face of shifting weather patterns.

Historically, traditional crop varieties have demonstrated remarkable adaptability to local environmental conditions, owing to the diverse genetic resources that have enabled farmers to navigate gradual climatic changes over the past 10,000 years (Gupta 2010). However, the rapid pace of climate change now poses a unique challenge in maintaining the optimal match between crops and their growing environments. In this context, crop wild relatives have emerged as a crucial resource, as their genetic diversity can be leveraged to impart resistance to the various environmental stressors associated with climate change (Gupta 2010; Bohra et al. 2020).

The development of new, climate-resilient crop varieties requires a multifaceted approach, harnessing the latest advancements in genomics and reverse genetics (Bohra et al. 2020; Mir et al. 2022). High-throughput phenotyping platforms and comprehensive crop models have enabled researchers to rapidly explore genetic resources, accelerating the discovery of beneficial traits and enhancing their physiological understanding of how crop improvement can increase tolerance to environmental stress (Mir et al. 2022). These advanced breeding techniques have been instrumental in the development of crop varieties that can thrive under unfavorable climatic conditions, ensuring reliable yields and safeguarding food security even as the climate continues to change.

At the same time, the successful adoption and scaling of climate-resilient crop varieties in developing countries face various institutional, environmental, socio-economic, and technical constraints. Overcoming these challenges will require a concerted, interdisciplinary effort, integrating advancements in plant breeding with a nuanced understanding of the local context and the needs of smallholder farmers. By harnessing the power of genomic approaches and ensuring their equitable deployment, the scientific community can play a pivotal role in empowering farmers to adapt to the impacts of climate change and secure a sustainable food future (Steenwerth et al. 2014; Kumari 2017; Singh et al. 2020).

3.1.5 Drought-Resistant Varieties

In regions prone to water scarcity and erratic rainfall patterns, cultivating crop varieties that are drought-tolerant is essential. Drought-tolerant crop varieties have developed mechanisms to survive and thrive under limited water conditions, ensuring that crops can withstand periods of drought stress without compromising yield potential. By incorporating drought-resistant traits into crop breeding programs, farmers can minimize crop losses and maintain productivity in water-stressed environments. Commonly in breeding for drought tolerance, grain yield is the basis for selection, but it is a complex, late-stage trait, affected by many factors aside from drought (Khadka et al. 2020). A strategy that evaluates genotypes for physiological responses to drought at earlier growth stages may be more targeted to drought and time efficient. Such an approach may be enabled by recent advances in high-throughput phenotyping platforms. This would allow for a more comprehensive understanding of the underlying mechanisms that confer drought tolerance, beyond just yield performance (Budak et al. 2013).

The contribution of selection based on physiological traits for drought adaptation has been small, and genetic engineering has at best contributed to incremental progress rather than causing a shift in tendencies. Drought-tolerant transgenic plants released in the last 10 years are not numerous, and have caused yield increases of 2–4%, i.e. within the same range as novel varieties obtained by conventional breeding. Novel breeding techniques such as genomic selection may well reinforce the power of yield-based selection, perhaps at the expense of physiology-based breeding.

Furthermore, it has been proposed that breeding strategies based on traits or genes for tolerance should only complement direct selection for yield, which would continue as the benchmark for all other strategies (Tardieu et al. 2018). Drought tolerance in modern and wild wheat has been a topic of increasing interest, as plants exhibiting high drought tolerance are the most suitable targets for drought-related research and are the most promising sources of drought-related genes and gene regions to be used in the improvement of modern crop varieties (Budak et al. 2013). The natural progenitors of cultivated crops, such as the goat grass (*Aegilops tauschii*), which is more drought-tolerant than *Triticum*, and wild emmer wheat, which harbors drought tolerance characteristics lost during the cultivation of modern lines, are of great importance (Budak et al. 2013). While development of higher-yielding crops under water-limited environments is the most viable solution to stabilizing and increasing wheat production under current climatic conditions, it is challenged by the nature of drought response as a trait and the complex genomic constitution of wheat (Budak et al. 2013). However, the utilization of drought-tolerant wild species and the rapid advances in molecular biological, functional genomics, and transgenics technologies have facilitated the breeding of drought-tolerant wheat varieties (Budak et al. 2013; Tardieu et al. 2018; Khadka et al. 2020; Kumar et al. 2023). For example, a recent study reported the development of a drought-tolerant bread wheat (*Triticum aestivum* L) variety HD3411 following marker-assisted backcrossing, which incorporated several physiological and morphological traits from a wild relative (Budak et al. 2013). This highlights the potential of leveraging wild germplasm to enhance drought tolerance in modern wheat cultivars.

3.1.6 Heat-Tolerant Varieties

The impacts of climate change, particularly the rise in global temperatures, pose significant challenges for agricultural production worldwide. As heatwaves and prolonged periods of high temperatures become more frequent, traditional crop varieties struggle to maintain growth and productivity (Yumurtaç 2015; Akter and Islam 2017; Zhao et al. 2022). To mitigate these adverse effects, the development of heat-tolerant crop varieties has emerged as a crucial strategy for ensuring consistent harvests in a warming climate.

By leveraging cutting-edge omics methods and speed breeding techniques, plant breeders can revolutionize the process of developing temperature-smart plants that can withstand fluctuations in temperature (Raza et al. 2024). These advanced tools, including genomics, transcriptomics, proteomics, metabolomics, and phenomics, provide a promising pathway to creating climate-proof crop varieties that can thrive in the face of rising global temperatures (Steenwerth et al. 2014; Kaushal et al. 2016; Akter and Islam 2017; Raza et al. 2024).

One of the key adaptive mechanisms that plants employ to cope with heat stress is the synthesis of heat-shock proteins, which play a vital role in maintaining protein structure and function under elevated temperatures. Similarly, the production of compatible solutes, such as sugars, amino acids, and betaines, helps plants regulate osmotic balance and protect cellular membranes during heat stress (Niu and Xiang

2018). Breeding efforts focused on enhancing the expression of genes responsible for the production of these protective molecules can significantly improve heat tolerance in food crops (Kaushal et al. 2016).

The development of transgenic plants overexpressing the enzymes involved in the synthesis of heat-protective compounds also holds promise for increasing thermotolerance in agricultural systems. These genetic approaches, when combined with the power of machine learning and speed breeding data, can revolutionize the creation of resilient plant varieties capable of withstanding the challenges posed by a changing climate (Kaushal et al. 2016; Raza et al. 2024).

By leveraging the latest advancements in omics-driven plant breeding, researchers can gain deeper insights into the physiological mechanisms underlying plant adaptation to heat stress. Analyzing gene expression patterns, protein profiles, and metabolite levels can help identify key genetic and molecular markers associated with heat tolerance, allowing for more targeted breeding strategies (Raza et al. 2024).

Additionally, the activation of stress-responsive transcription factors, which regulate the expression of genes involved in various adaptive pathways, offers another avenue for enhancing heat tolerance in crops (Kaushal et al. 2016; Raza et al. 2024). Overexpressing these transcription factors or introducing them through genetic manipulation can confer improved resilience to elevated temperatures (Shukla and Mattoo 2013; Kaushal et al. 2016; Niu and Xiang 2018). For example, studies have shown that the over-expression of heat shock transcription factors in plant species such as rice, wheat, and maize can lead to enhanced thermotolerance, enabling these crops to maintain growth and productivity even under heat stress (Kaushal et al. 2016). Furthermore, the identification and utilization of stress-tolerant wild plant species as models can provide valuable insights into the mechanisms underlying abiotic stress adaptation, which can then be applied to the genetic improvement of cultivate.

By incorporating these advanced breeding techniques and genetic engineering approaches, researchers can develop a new generation of heat-tolerant crop varieties that are better equipped to withstand the challenges of a warming climate (Boscaiu Neagu et al. 2012; Shukla and Mattoo 2013; Kaushal et al. 2016; Niu and Xiang 2018). This multifaceted approach, leveraging the latest scientific advancements, is crucial for ensuring food security and supporting sustainable agricultural practices in the face of climate change.

3.1.7 Pest-Resistant Varieties

Pests and diseases pose significant threats to agricultural crops, leading to yield losses and compromising food security. Climate change can exacerbate pest pressures by altering pest distribution patterns and increasing pest outbreaks (Zhan et al. 2015). Developing pest-resistant crop varieties that possess natural resistance to common pests and diseases is a promising approach to sustainable pest management. By incorporating natural resistance mechanisms into crop plants, farmers can reduce their reliance on chemical pesticides while effectively protecting their crops from devastating pest damage (Zhan et al. 2015; Velasco et al. 2020).

Pest-resistant crop varieties contribute to a more holistic and resilient approach to agricultural production. Through the strategic development of crops with inherent resistance to pests and diseases, the industry can mitigate the risks posed by evolving pathogen populations and minimize the environmental consequences of excessive pesticide use. Modern agricultural practices have increasingly relied on agrochemicals and formulations for disease control, which can lead to environmental pollution and have detrimental effects on human health (Bhoi et al. 2022). As the global human population increases and arable land declines, there is an urgent need to explore alternative, sustainable pest management strategies to ensure food security (Popp et al. 2013). Genetic improvement through the development of pest-resistant crop varieties is a promising approach that can enhance the resilience of agricultural systems in the face of evolving pest challenges (Bhoi et al. 2022).

3.1.8 Integrated Pest Management

Integrated Pest Management (IPM) has emerged as a comprehensive and environmentally conscious approach to tackle pest challenges in agricultural systems. At the core of IPM is the principle of leveraging ecological processes to manage pest populations, reducing the reliance on chemical pesticides. A key aspect of IPM is the utilization of biological control, which involves harnessing the power of natural predators, parasites, and pathogens to regulate pest populations. By encouraging the presence and proliferation of beneficial organisms, IPM strategies enhance the ecosystem's natural checks and balances, leading to more sustainable and effective pest suppression. Crop rotation, another central practice in IPM, disrupts the life cycles of pests and limits their ability to establish continuous populations. This approach, combined with continuous monitoring of pest levels, enables farmers to implement proactive and targeted control measures before pest problems escalate, minimizing the need for broad-spectrum pesticide applications. The integration of these ecological strategies within IPM not only reduces the reliance on synthetic pesticides but also mitigates the environmental impact of pest control, promoting a more balanced and resilient agroecosystem.

Biopesticides, as a component of IPM, have gained increasing attention in recent years as an environmentally friendly alternative to conventional chemical pesticides. These naturally derived agents, such as microbial organisms and plant-based compounds, can effectively manage pests while posing a lower risk to human health and the environment. The use of biopesticides in IPM programs complements physical and cultural control methods, allowing for a more holistic and integrated approach to crop protection.

Studies have shown that biopesticides can be valuable components of IPM, promoting sustainable agriculture by leveraging natural mechanisms of pest regulation. Biopesticides, which include microbial organisms and plant-based compounds, have gained significant interest in recent years as an environmentally friendly alternative to conventional chemical pesticides. Their integration into IPM strategies can provide a more harmonious approach to pest management, complementing physical and cultural control methods to minimize the need for synthetic pesticides.

The adoption of IPM practices, incorporating both biological control and biopesticides, has the potential to transform the landscape of agricultural pest management. By harnessing ecological principles and reducing the reliance on chemical inputs, IPM offers a path toward more sustainable and resilient agricultural systems (Campos et al. 2016).

Biological control, a cornerstone of IPM, leverages the natural predators, parasites, and pathogens within ecosystems to regulate pest populations (Campos et al. 2016). These natural enemies can effectively suppress pests, reducing the need for chemical interventions and contributing to a more balanced agroecosystem. Crop rotation, another key practice in IPM, disrupts the life cycles of pests, limiting their ability to establish continuous populations. Furthermore, the continuous monitoring of pest levels enables proactive and targeted control measures, minimizing the need for broad-spectrum pesticide applications and promoting a more precise and efficient approach to pest management.

A ladybird beetle (*Coccinellidae*) consuming aphids on a bean plant

3.2 Soil

Soil is a fundamental component of agroecology, and its management plays a crucial role in sustainable agricultural practices. Agroecology recognizes the importance of soil health and emphasizes strategies that promote soil conservation, fertility, and biodiversity.

3.2.1 Soil Health

Agroecology places a strong emphasis on maintaining and enhancing soil health as the foundation for sustainable agriculture. Healthy soils are characterized by a range of key attributes, including good physical structure, adequate nutrient content, high organic matter levels, and diverse microbial communities (Guo 2021). Agroecological systems employ a variety of practices, such as organic farming, cover cropping, and reduced tillage, to improve and preserve soil health (Gliessman 1995). Organic farming, a central component of agroecology, aims to minimize the use of synthetic inputs and prioritize the use of organic matter and natural processes (Coyne et al. 2022). This approach promotes soil health by enhancing soil structure, nutrient cycling, and microbial activity. Research has shown that organic farming systems tend to have higher soil organic matter content, better soil structure, and increased microbial biomass compared to conventional systems, highlighting the positive impacts of this practice (Gliessman 1995).

Cover cropping is another key technique used in agroecological systems to improve soil health. Cover crops are planted during fallow periods or alongside cash crops to cover and protect the soil. They contribute to soil health by adding organic matter, improving soil structure, increasing nutrient availability, and suppressing weeds. Research has demonstrated that cover cropping practices in agroecological systems significantly increase soil organic matter content and overall soil quality (Swami 2020; Hsiao et al. 2017). Reduced tillage, or conservation tillage, is another practice commonly employed in agroecological systems to minimize soil disturbance. This technique helps preserve soil structure, reduce erosion, enhance water infiltration, and promote the accumulation of organic matter in the soil. Studies have shown that reduced tillage practices in agroecology can lead to increased soil organic matter levels and improved soil health compared to conventional tillage methods.

The combination of these agroecological practices, including organic farming, cover cropping, and reduced tillage, works synergistically to enhance and maintain soil health, which is the foundation for sustainable and productive agricultural systems. Organic farming, for instance, has been shown to increase soil organic matter, improve soil structure, and promote microbial diversity compared to conventional farming practices (Swami 2020). Similarly, cover cropping and reduced tillage have been found to enhance soil organic matter, nutrient cycling, and overall soil quality

(Key et al. 2016; Degrune et al. 2019; Swami 2020; Sethuraman et al. 2021). Ultimately, the adoption of agroecological approaches that prioritize soil health can lead to more resilient and environmentally friendly farming systems that can meet the challenges of modern agriculture.

3.2.2 Soil Fertility

Agroecology places a strong emphasis on the utilization of natural and organic sources of nutrients to enhance soil fertility in agricultural systems. This approach recognizes the crucial role of organic matter additions, such as compost and manure, in providing a steady supply of essential nutrients, improving soil structure, and promoting beneficial microbial activity (Nayak et al. 2020; Swami 2020). Organic matter additions contribute to the overall nutrient balance of the soil, supporting plant growth and promoting sustainable agricultural practices. A study by Swami (2020) demonstrated that the addition of compost and manure in agroecological systems significantly improved soil fertility and nutrient content (). In addition to nutrient supply, organic matter additions in agroecology improve soil structure and promote beneficial microbial activity. Organic materials contribute to the formation of stable aggregates, enhancing soil porosity and water infiltration, while also reducing erosion risks. They also serve as a substrate for beneficial microorganisms, such as bacteria and fungi, which play a vital role in nutrient cycling and plant nutrient uptake.

Crop rotation is another essential practice in agroecology that helps enhance nutrient cycling and reduce nutrient imbalances. By alternating the cultivation of different crops in sequence, crop rotation diversifies nutrient demands, preventing the depletion of specific nutrients and fostering a more balanced nutrient profile in the soil (Dotaniya et al. 2020; Nayak et al. 2020; Swami 2020). This, in turn, can support the growth of a wider range of crops, enhancing the overall productivity and resilience of the agroecological system. Crop rotation is particularly beneficial in agroecological systems as it helps to break pest and disease cycles, while also contributing to the maintenance of soil organic matter and nutrient levels.

3.2.3 Soil Conservation

Agroecological practices prioritize soil conservation as a fundamental aspect of sustainable agriculture, focusing on preventing erosion, maintaining soil structure, and minimizing nutrient losses. Among the key techniques employed in agroecology to mitigate soil erosion are contour plowing, terracing, and conservation tillage, which aim to minimize soil disturbance and preserve soil cover (Oyana et al. 2015; Tamburini et al. 2016; Blanchy et al. 2023). Agroforestry systems and buffer strips also serve as vegetative barriers to further contribute to soil conservation by filtering runoff and stabilizing soils.

Contour plowing, a soil conservation practice fundamental to agroecology, involves plowing perpendicular to the slope of the land rather than up and down the slope. This technique helps reduce the speed and volume of runoff, allowing more water to infiltrate the soil and mitigating erosion.

Contour farming in Machakos County, Kenya

Terracing is another agroecological strategy used to prevent soil erosion on steep slopes. Terraces are horizontal platforms constructed across the slope, creating flat areas for cultivation. By reducing the slope length, terraces help slow down runoff and promote infiltration, thereby curbing erosion. A study by Zhang et al. (2021) found that terracing significantly reduced soil erosion and improved soil conservation on steep agricultural slopes.

Conservation tillage, a key soil management practice in agroecology, aims to minimize disturbance of the soil. This approach involves reducing or eliminating plowing, leaving crop residues on the surface, and using specialized implements that disturb the soil minimally. Conservation tillage helps maintain soil structure, increase water infiltration, and reduce erosion by preserving the soil's natural protections (Tamburini et al. 2016; Degrune et al. 2019; Ojo et al. 2023).

The use of agroforestry systems and buffer strips in agroecological settings further contributes to soil conservation efforts. These vegetative barriers serve to filter runoff and stabilize soils, mitigating the impacts of erosion. Farmers who successfully implement soil conservation methods experience improved production and livelihoods (Ojo et al. 2023). Overall, the suite of agroecological practices focused on soil conservation, such as contour plowing, terracing, and conservation tillage, work in tandem to promote the long-term health and productivity of agricultural lands.

3.2.4 Soil Biodiversity

Agroecology recognizes the pivotal role of soil biodiversity in sustaining ecosystem functioning and resilience within agricultural systems. Diverse farming practices, such as crop rotations, intercropping, and agroforestry, are implemented in agroecology to promote soil biodiversity by creating habitats for a wide range of organisms, including beneficial insects, earthworms, and microorganisms (Dighton and Krumins 2014). Soil biodiversity plays a vital role in nutrient cycling, pest regulation, and overall soil health, which are essential for the long-term sustainability of agricultural systems (Altieri 1999). Crop rotations, a common practice in agroecology, have been shown to enhance soil biodiversity by providing varied root exudates, organic matter inputs, and microhabitats for different organisms. The positive impact of crop rotations on soil biodiversity is associated with increased abundance and diversity of beneficial soil organisms, including earthworms, beneficial nematodes, and microbial communities (Jat et al. 2021). Intercropping, the cultivation of two or more different crops simultaneously in the same field, also contributes to soil biodiversity in agroecological systems (Jat et al. 2021). The diversity of plant species and root systems in intercropping systems provides a range of resources and niches for soil organisms, promoting their abundance and diversity.

Research by Jat et al. (2021) demonstrated that intercropping systems in agroecology enhanced soil microbial diversity, nutrient cycling, and overall soil health compared to monocropping systems. Agroforestry, which integrates trees with crops or livestock, plays a significant role in promoting soil biodiversity as well. Trees provide habitat, food sources, and microclimatic conditions that support diverse soil organisms, including beneficial insects, mycorrhizal fungi, and decomposers. A study by Eisenhauer et al. (2018) emphasizes the positive impact of agroforestry on soil biodiversity, with increased microbial biomass, earthworm abundance, and insect diversity. The diverse agricultural practices implemented in agroecology, such as crop rotations, intercropping, and agroforestry, create a mosaic of habitats and resources that foster the proliferation of a wide range of soil-dwelling organisms (Altieri 1999; Degrune et al. 2019; Jat et al. 2021). This biodiversity-friendly approach promotes the ecological functions that are essential for sustainable agricultural production, including nutrient cycling, pest regulation, and soil health maintenance.

3.2.5 Carbon Sequestration

Agroecology, with its emphasis on sustainable and ecological farming practices, plays a crucial role in climate change mitigation through its potential for carbon sequestration in soils. One of the key agroecological practices that contributes to this carbon sequestration is cover cropping. Cover crops, typically planted during fallow periods or between main crop cycles, provide a living cover that protects the soil from erosion and enhances organic matter inputs. As cover crops grow, they accumulate carbon through photosynthesis and deposit it into the soil upon decomposition. A study by Amelung et al. (2020) demonstrated that cover cropping increased soil organic carbon stocks and enhanced carbon sequestration potential in agricultural soils.

Another agroecological practice that contributes to carbon sequestration in soils is crop residue retention. Instead of removing crop residues, such as stalks and leaves, after harvest, they are left on the soil surface as organic mulch. Crop residues gradually decompose, releasing carbon into the soil and contributing to soil organic carbon accumulation, as highlighted by research from Morgan et al. (2010).

Agroforestry systems, with their combination of trees and crops or livestock, also have considerable potential for carbon sequestration. Trees in agroforestry systems sequester carbon through photosynthesis and store it in above-ground biomass, including leaves, stems, and branches. The roots of trees also contribute to soil carbon storage, and the leaf litter and organic matter produced by trees further enhance organic carbon inputs to the soil (Morgan et al. 2010; Mayer et al. 2018; Amelung et al. 2020; Liu et al. 2024). The implementation of these agroecological practices, including cover cropping, crop residue retention, and agroforestry, can be an effective strategy for climate change mitigation by increasing carbon sequestration in agricultural soils (Morgan et al. 2010; Liu et al. 2024). The carbon farming solution: a global toolkit of perennial crops and regenerative agriculture practices for climate change mitigation and food security (Mayer et al. 2018).

Crop production systems and their interactions with the environment, known as agroecosystems, cover about one-third of the Earth's land surface (Liu et al. 2024). As soil constitutes the largest single carbon reservoir on land, agroecosystems play a key role in the global terrestrial carbon cycle through crop interactions with soils and atmosphere (Liu et al. 2024). Globally, agriculture is a significant source of greenhouse gases; yet, carbon uptake by crops also removes large amounts of carbon dioxide from the atmosphere, some of which can be stabilized in soil. Because most intensively cultivated soils are carbon-unsaturated, practices that increase soil organic carbon represent a low-cost, large-scale strategy for reducing atmospheric greenhouse gas concentrations (Mayer et al. 2018). Nutrient availability, biomass productivity, the type of vegetation, and water availability can be major limiting factors inhibiting increases in soil carbon sequestration (Zomer et al. 2017). Nonetheless, the imperative to sequester carbon in soils arises from the multiple co-benefits that are obtained from sequestration of carbon in soils that have been depleted of their organic matter, such as improved soil fertility, health, and functioning (Zomer et al. 2017; Liu et al. 2024).

Toward a global-scale soil carbon mitigation strategy, sustainable soil carbon sequestration practices need to be rapidly scaled up and implemented. The major potential for carbon sequestration is in cropland soils, especially those with large yield gaps and/or large historic soil organic carbon losses (Zomer et al. 2017; Amelung et al. 2020). The establishment of a soil information system containing localized information on soil group, degradation status, crop yield gap, and the associated carbon sequestration potentials, as well as the provision of incentives and policies to translate management options into region- and soil-specific practices, can facilitate the implementation of effective soil carbon sequestration measures (Zomer et al. 2017; Mayer et al. 2018; Amelung et al. 2020; Liu et al. 2024).

3.2.6 Soil Water Management

Agroecological practices play a crucial role in optimizing water-use efficiency and enhancing soil water-holding capacity to promote sustainable agriculture. These practices, such as mulching, agroforestry, and contour plowing, contribute to reducing water runoff, increasing infiltration, and minimizing soil moisture losses, thereby improving resilience to drought conditions and enhancing water availability for crops (Stroosnijder et al. 2012).

Mulching, a common practice in agroecology, involves covering the soil surface with a layer of organic or inorganic materials, such as crop residues, straw, or plastic sheets. Mulches act as a protective layer, reducing evaporation, preventing direct exposure of the soil to sunlight, and minimizing water loss through runoff. Research has demonstrated that mulching significantly reduced soil moisture evaporation and improved water-use efficiency in agricultural systems (He et al. 2018). Agroforestry systems, which integrate trees with crops or livestock, play a crucial role in enhancing water availability. The presence of trees in agroforestry systems contributes to increased infiltration and reduced runoff by intercepting rainfall, reducing its impact on the soil surface. Trees also contribute to the creation of microclimates that promote water conservation and reduce evaporation. Studies have highlighted the positive impact of agroforestry on water availability, emphasizing increased water infiltration and reduced runoff in agroforestry systems compared to conventional agriculture (He et al. 2018; Ashok et al. 2020). Contour plowing, another soil conservation technique employed in agroecology, helps optimize water-use efficiency. By plowing along the contour lines of the land, contour plowing slows down water flow, facilitates water infiltration, and reduces soil erosion. This practice helps retain water within the root zone and enhances soil water availability for crops. Research has demonstrated the effectiveness of contour plowing in optimizing water-use efficiency.

Plastic mulch applied to crops within a dryland agroecological zone

The integration of these agroecological practices, such as mulching, agroforestry, and contour plowing, can significantly improve water-use efficiency and enhance the resilience of agricultural systems to drought conditions (He et al. 2018; Ashok et al. 2020; Kumawat et al. 2020). Mulching, for instance, has been shown to reduce soil moisture evaporation and improve water-use efficiency by acting as a protective layer that prevents direct exposure of the soil to sunlight and minimizes water loss through runoff (Ashok et al. 2020). Agroforestry systems, on the other hand, contribute to increased infiltration and reduced runoff by intercepting rainfall, while also creating microclimates that promote water conservation and reduce evaporation. Contour plowing, a soil conservation technique, further enhances water-use efficiency by slowing down water flow, facilitating water infiltration, and reducing soil erosion, thereby helping to retain water within the root zone and increasing soil water availability for crops (Nair 2007; Ong et al. 2014; Colmenares et al. 2020).

The effectiveness of these agroecological practices in optimizing water management has been well-documented in the scientific literature. A reduction in soil moisture evaporation and improved water-use efficiency are achieved through mulching in agricultural systems. Research by Nair (2014) highlighted the positive impact of agroforestry on water availability, with increased infiltration and reduced runoff observed in agroforestry systems compared to conventional agriculture. Furthermore, the effectiveness of contour plowing in optimizing water-use efficiency through its ability to slow down water flow facilitates infiltration and reduces soil erosion (He et al. 2018).

The integration of these water-conserving agroecological practices can significantly improve the resilience of agricultural systems to drought conditions. By reducing water runoff, increasing infiltration, and minimizing soil moisture losses, these practices can enhance water availability for crops, thereby promoting sustainable agriculture and ensuring food security even in the face of increasing water scarcity (Stroosnijder et al. 2012). The strategic combination of mulching, agroforestry, and contour plowing can deliver synergistic benefits, optimizing water-use efficiency and strengthening the capacity of agricultural systems to adapt to climate change (He et al. 2018; Ashok et al. 2020).

3.3 Natural Vegetation

Natural vegetation plays a crucial role in agroecology as it serves as a reference point and inspiration for designing sustainable agricultural systems. Agroecology aims to mimic the diversity, structure, and functions of natural ecosystems to enhance agricultural productivity, biodiversity, and ecological resilience.

3.3.1 Biodiversity Conservation

Natural vegetation plays a crucial role in supporting biodiversity, including native plant species, beneficial insects, pollinators, and other wildlife. Agroecology recognizes the significance of preserving and enhancing biodiversity within agricultural landscapes. By incorporating elements of natural vegetation into agricultural systems, such as hedgerows, buffer strips, and cover crops, agroecological practices provide habitats and resources for diverse species, contributing to ecological balance and pest regulation. Hedgerows are linear strips of vegetation composed of trees, shrubs, and herbaceous plants that are commonly integrated into agroecological systems. Hedgerows serve as wildlife corridors, providing shelter, nesting sites, and foraging opportunities for a variety of species, including birds, beneficial insects, and small mammals (Altieri 1999). Hedgerows significantly enhance biodiversity by increasing species richness and abundance of beneficial insects in agricultural landscapes (Altieri 1999). Buffer strips are areas of vegetation, typically grasses or wildflowers, planted alongside agricultural fields or water bodies. These strips act as vegetative barriers, protecting against soil erosion, filtering runoff, and providing habitat for wildlife. Buffer strips promote biodiversity by creating corridors for wildlife movement, supporting pollinators, and fostering the presence of beneficial insects that contribute to natural pest control. Research by Altieri (1999) and Batáry et al. (2011) highlight the positive impact of buffer strips on biodiversity, emphasizing increased species richness and abundance of beneficial insects in agricultural landscapes.

Cover crops are non-cash crops grown between main crop cycles or during fallow periods. Cover crops contribute to biodiversity in agroecological systems by providing additional habitats and resources for a variety of species. The plant diversity and structure of cover crops attract beneficial insects, which can help regulate pest populations and support ecosystem balance (Altieri 1999; Gaba et al. 2015; Degrune et al. 2019).

3.3.2 *Ecosystem Services*

Natural vegetation plays a crucial role in providing a range of ecosystem services that are essential for agricultural productivity, including nutrient cycling, soil erosion control, water regulation, pollination, and biological pest control. Agroecological practices aim to harness and enhance these ecosystem services through techniques such as agroforestry, intercropping, and the establishment of ecological corridors. By integrating natural vegetation elements into agricultural systems, agroecology enhances ecosystem services and reduces reliance on external inputs (Power 2010).

Nutrient cycling is a fundamental ecosystem service facilitated by natural vegetation. Native plants and diverse vegetation contribute to nutrient cycling by extracting and redistributing nutrients from the soil through their root systems, helping maintain nutrient availability and fertility in the agroecosystem (Garibaldi et al. 2018). The presence of natural vegetation enhances nutrient cycling, resulting in improved nutrient availability and crop productivity (Garibaldi et al. 2018).

Dung beetles feeding on a cow's dropping. These are efficient ecosystem engineers recycling nutrients

Soil erosion control is another critical ecosystem service provided by natural vegetation. The roots of plants hold the soil in place, preventing erosion caused by wind and water, and vegetation cover, such as grasses and trees, reduces the impact of rainfall, minimizing soil runoff and erosion. Studies by Lin (2011) and Power (2010) underscore the importance of natural vegetation in preventing soil erosion and maintaining soil health.

Water regulation is also facilitated by natural vegetation through its ability to regulate water flow, enhance water infiltration, and reduce water runoff. Vegetation cover acts as a natural sponge, absorbing rainfall and releasing it slowly into the soil, reducing the risk of flooding and enhancing water availability during dry periods. Research by Daily et al. (2007) and Power (2010) highlight the positive impact of natural vegetation on water regulation, emphasizing increased water infiltration rates.

In addition to these vital services, natural vegetation plays a crucial role in supporting biodiversity and providing habitat for beneficial organisms. Non-crop habitats, such as those with natural vegetation, provide the resources required for arthropod predators, parasitoids, insectivorous birds and bats, and microbial pathogens that act as natural enemies to agricultural pests, thereby providing biological control services in agroecosystems (Power 2010). These biological control services can reduce the need for pesticides, as highlighted by a study that estimated the value of natural enemy suppression of soybean aphid in the United States to be a minimum of $239 million in 2007–2008 alone (Power 2010).

The importance of natural vegetation in agroecosystems extends beyond the direct provision of ecosystem services. A growing body of research indicates that the greater functional capacity of diverse agroecosystems, which incorporate natural vegetation, helps protect crop productivity against environmental changes (Myers 1996; Lin 2011). The resilience and adaptability of these systems are crucial in the face of expected biotic and abiotic shifts due to climate change and other factors. Through the integration of natural vegetation into agricultural systems, agroecological practices harness the full potential of these vital ecosystem services, promoting sustainable and resilient food production.

3.3.3 Soil Health and Fertility

Natural vegetation plays a pivotal role in contributing to soil health and fertility through the promotion of organic matter accumulation, nutrient cycling, and soil structure improvement (Jose 2009; Stokes et al. 2014; Fontaine et al. 2024). Agroecology draws inspiration from natural ecosystems and incorporates practices such as cover cropping, green manure, and crop rotation to enhance soil fertility and structure (Nayak et al. 2020; Swami 2020). By emulating the dynamics of natural vegetation, agroecological systems can maintain long-term soil productivity and reduce reliance on synthetic fertilizers (Teague 2018; Nayak et al. 2020; Fontaine et al. 2024). Natural vegetation, with its diverse plant species and associated root systems, contributes to organic matter accumulation in the soil. Plant residues, including fallen leaves, stems, and roots, provide a continuous source of organic matter (Fontaine et al. 2024). This organic matter enhances soil fertility by improving nutrient content, water holding capacity, and promoting the growth of beneficial microorganisms (Swami 2020).

Nutrient cycling is facilitated by natural vegetation through the uptake and redistribution of nutrients from the soil. Plant roots absorb nutrients, and when plants die or shed their leaves, these nutrients are released back into the soil through decomposition processes. This cycling of nutrients contributes to maintaining nutrient availability and balance in the soil. Agroecological practices such as cover cropping and green manure utilize diverse plant species that enhance nutrient cycling by capturing and recycling nutrients (Nayak et al. 2020; Fontaine et al. 2024). Cover cropping and green manure use have positive impacts on nutrient cycling and soil fertility in agricultural systems (Nayak et al. 2020; Swami 2020; Fontaine et al. 2024). Soil structure improvement is another key benefit provided by natural vegetation. Plant roots penetrate the soil, creating channels and pores that improve soil structure, porosity, and water infiltration (Carranza-Gallego et al. 2018; Dotaniya et al. 2020; Nayak et al. 2020; Swami 2020). These improved soil physical properties enhance the ability of the soil to support plant growth and water retention, ultimately contributing to the long-term sustainability of agroecological systems.

3.3.4 Water Management

Natural vegetation, including forests, wetlands, and other types of natural habitats, plays a crucial role in regulating water availability and quality. These ecosystems act as natural reservoirs that contribute to water regulation, erosion prevention, and enhanced water infiltration. Agroecology recognizes the importance of integrating elements of natural vegetation into agricultural systems to improve water management. Practices such as agroforestry and the establishment of riparian buffers are employed to conserve water, reduce runoff, and enhance water quality (Ong et al. 2014). Forests are known for their exceptional ability to regulate water flow and maintain water availability (Nugroho et al. 2019). The dense canopy of trees intercepts rainfall, reducing the impact of raindrops on the soil surface and minimizing soil erosion. The root systems of trees also enhance water infiltration, allowing water to percolate into the soil and recharge groundwater reservoirs (Ong et al. 2014). Forests, thus, aid in regulating water availability and maintain streamflow mitigating water scarcity.

Wetlands are another valuable natural ecosystem that provides numerous water-related benefits, including water filtration and flood control. Wetlands act as natural filters, removing sediments, nutrients, and pollutants from water as it passes through their vegetation and soils. This process improves water quality and reduces the negative impacts of excessive nutrients and sedimentation on aquatic ecosystems. Agroecological practices integrate elements of natural vegetation into agricultural systems to enhance water management. Agroforestry, which combines trees with crops or livestock, helps regulate water flow by intercepting rainfall, reducing soil erosion, and enhancing water infiltration (Ong et al. 2014). The presence of trees within agricultural systems can also create a microclimate that reduces evaporation and maintains groundwater levels, especially during the dry season.

3.3.5 Resilience to Climate Change

Natural vegetation has demonstrated remarkable resilience in the face of climate variability and extreme events, offering valuable insights for agroecological practices (Cabell and Oelofse 2012). Agroecology draws upon the resilience and adaptive capacity of natural ecosystems to enhance the resilience of agricultural systems and mitigate the impacts of climate change (Altieri and Nicholls 2013). By promoting diverse crops, intercropping, and agroforestry, agroecological systems can increase their ability to withstand climatic stresses, including drought, pests, and diseases.

Diverse cropping systems are a key aspect of agroecology and contribute to enhancing resilience. Crop diversity provides a range of genetic resources that enable plants to adapt to changing environmental conditions, as demonstrated by studies showing that diverse cropping systems are more resilient to climate variability and extreme weather events (Cabell and Oelofse 2012). An analysis by Lin (2011) found that crop diversity positively influenced the stability and productivity of agroecosystems, reducing their vulnerability to climate-related risks.

Intercropping, the practice of growing two or more crops simultaneously in the same field, offers several benefits for climate resilience. By combining different crop species, intercropping enhances resource-use efficiency, reduces pest and disease pressures, and improves soil health. Intercropping systems can buffer the negative impacts of climate variability by spreading risks across different crops and increasing overall system stability, as observed in a study by Raseduzzaman and Jensen (2017) that found intercropping improved yield stability and reduced the vulnerability of agricultural systems to climatic fluctuations.

Agroforestry, which integrates trees with crops or livestock, plays a vital role in enhancing climate resilience in agricultural systems. Trees provide numerous benefits, including shade, windbreaks, and nutrient cycling, which help mitigate climate-related stresses (Altieri and Nicholls 2013; Degrune et al. 2019; Koskey et al. 2021). Agroforestry systems have been shown to improve soil moisture and fertility, enhance biodiversity, and increase the capacity of farming communities to adapt to the effects of climate change (Altieri and Nicholls 2013). The integration of trees and crops in agroforestry systems can diversify production, improve nutrient cycling, and provide a buffer against environmental fluctuations (Koskey et al. 2021).

Observations of agricultural performance after extreme climatic events have revealed that resilience to climate disasters is closely linked to high levels of on-farm biodiversity, a typical feature of traditional farming systems. Various experts have suggested that rescuing traditional management systems combined with the use of agroecologically based management strategies may represent a viable and robust path to increase the productivity, sustainability, and resilience of smallholder agricultural production under predicted climate scenarios (Altieri and Nicholls 2013; Koskey et al. 2021).

Ultimately, the resilience of agroecological systems is rooted in their ability to mimic the diversity, complexity, and adaptability of natural ecosystems. By drawing on the inherent strengths of natural vegetation, agroecological practices such as diverse cropping, intercropping, and agroforestry can enhance the capacity of agricultural systems to withstand the challenges posed by climate change (Nair 2007; Altieri and Nicholls 2013; Mubiru et al. 2018; Koskey et al. 2021). Diversifying crop species not only provides a range of genetic resources for adaptation but also improves resource-use efficiency, reduces pest and disease pressures, and enhances overall system stability. Intercropping, which combines different crops in the same field, has been shown to buffer the negative impacts of climate variability by spreading risks across multiple cultivars and improving yield stability. Similarly, the integration of trees into agroforestry systems can improve soil moisture and fertility, enhance biodiversity, and increase the capacity of farming communities to adapt to environmental fluctuations (Nair 2007).

3.3.6 Landscape Connectivity

Natural vegetation plays a critical role in providing connectivity and corridors for species movement and gene flow within ecosystems (Saura and Pascual-Hortal 2007). Agroecology acknowledges the significance of landscape connectivity for biodiversity conservation and the maintenance of essential ecosystem functions. By incorporating elements of natural vegetation, such as wildlife corridors and habitat patches, into agricultural landscapes, agroecological systems actively contribute to ecological connectivity and the support of wildlife populations. Wildlife corridors are linear landscape features that connect fragmented habitats, allowing for the movement of species between different areas. These corridors promote gene flow, genetic diversity, and species resilience by facilitating migration, dispersal, and colonization (Saura and Pascual-Hortal 2007). Research has demonstrated the positive impacts of wildlife corridors on species movements and emphasized their importance in maintaining ecological connectivity and mitigating the negative effects of habitat fragmentation (Saura and Pascual-Hortal 2007; Batáry et al. 2011). Habitat patches within agricultural landscapes, which consist of natural vegetation elements like hedgerows, buffer strips, and cover crops, provide local habitats for a variety of plant and animal species. These patches offer shelter, food resources, and breeding grounds for wildlife, contributing to biodiversity conservation within agroecological systems (Estrada-Carmona et al. 2022). A growing body of research has highlighted the role of habitat patches in enhancing biodiversity and promoting the presence of beneficial insects, pollinators, and other wildlife in agricultural landscapes (Batáry et al. 2011; Estrada-Carmona et al. 2022).

Agroecology aims to learn from and collaborate with natural ecosystems to develop sustainable agricultural systems. By integrating the principles and functions of natural vegetation, agroecology strives to enhance biodiversity, ecosystem services, soil health, water management, and climate resilience within agricultural landscapes. This approach recognizes the interconnectedness of agricultural systems with the surrounding environment, and the critical role that natural vegetation plays in supporting the overall health and resilience of these systems (Gliessman 1995). Ecologists and conservation practitioners have often viewed agricultural land as a biological desert or a hostile matrix isolating habitat fragments. However, agroecosystems may also provide environmental benefits and important ecosystem services, such as pollination and biological control (Batáry et al. 2011).

The holistic philosophy of agroecology maintains the concept of connectivity and interdependence at the forefront, influencing the development of communities and the agroecology systems needed to sustainably support them. At the landscape scale, agroecology is the science and practice of sustainable agriculture based on ecological principles that operated on a landscape scale (Winter et al. 2020).

The integration of natural vegetation elements, such as wildlife corridors and habitat patches, into agricultural landscapes is a key aspect of agroecological design. These landscape features promote connectivity, allowing the movement of species between fragmented habitats and supporting the maintenance of genetic diversity and ecosystem functions. By fostering these connections, agroecological systems can contribute to the conservation of biodiversity and the overall resilience of agricultural landscapes.

The restoration and enhancement of wildlife corridors within agricultural areas can have significant benefits for species movement and gene flow. For example, Haddad et al. (2015) demonstrated the positive impacts of wildlife corridors on species dispersal and emphasized their importance in mitigating the negative effects of habitat fragmentation. Similarly, habitat patches composed of natural vegetation elements, such as hedgerows and buffer strips, can provide critical resources and breeding grounds for a variety of plant and animal species within agroecological landscapes (Degrune et al. 2019). Research by Batáry et al. (2011) and Degrune et al. (2019) highlight the role of these habitat patches in enhancing biodiversity and supporting the presence of beneficial insects, pollinators, and other wildlife.

References

Abbas F, Hammad HM, Fahad S, Cerdà A, Rizwan M, Farhad W, Ehsan S, Bakhat HF (2017, March 9) Agroforestry: a sustainable environmental practice for carbon sequestration under the climate change scenarios—a review. Environ Sci Pollut Res 24(12):11177–11191. https://doi.org/10.1007/s11356-017-8687-0

Akter N, Islam M (2017) Heat stress effects and management in wheat: a review. Agron Sustain Dev 37(5). https://doi.org/10.1007/s13593-017-0443-9

Altieri MA (1999) The ecological role of biodiversity in agroecosystems. Agric Ecosyst Environ 74(1–3):19–31. https://doi.org/10.1016/s0167-8809(99)00028-6

Altieri MA, Nicholls CI (2013) The adaptation and mitigation potential of traditional agriculture in a changing climate. Clim Chang 140(1):33–45. https://doi.org/10.1007/s10584-013-0909-y

Amelung W, Bossio D, Vries WD, Kögel-Knabner I, Lehmann J, Amundson R et al (2020) Towards a global-scale soil climate mitigation strategy. Nat Commun 11(1):5427. https://doi.org/10.1038/s41467-020-18887-7

Ashok A, Arunadevi K, Sujitha E (2020) Improving water use efficiency in greenhouse cultivation: a review. JCS 8(5):543–546. https://doi.org/10.22271/chemi.2020.v8.i5h.10351

Bakker T, Wouters H, van Asselt K, Bontsema J, Tang L, Müller J, van Straten G (2008) A vision based row detection system for sugar beet. Comput Electr Agric (CEA) 60(1):87–95. https://doi.org/10.1002/rob.21995

Batáry P, Báldi A, Kleijn D, Tscharntke T (2011) Landscape-moderated biodiversity effects of agri-environmental management: a meta-analysis. Proc R Soc B Biol Sci 278(1713):1894–1902. https://doi.org/10.1098/rspb.2010.1923

Bhoi A, Yadu B, Chandra J, Keshavkant S (2022) Mutagenesis: a coherent technique to develop biotic stress resistant plants. Plant Stress 3:100053. https://doi.org/10.1016/j.stress.2021.100053

Blanchy G, Bragato G, Di Bene C, Jarvis N, Larsbo M, Meurer K, Garré S (2023) Soil and crop management practices and the water regulation functions of soils: a qualitative synthesis of meta-analyses relevant to European agriculture. Soil 9(1):1–20. https://doi.org/10.5194/soil-9-1-2023

Bohra A, Chand Jha U, Godwin ID, Kumar Varshney R (2020) Genomic interventions for sustainable agriculture. Plant Biotechnol J 18(12):2388–2405. https://doi.org/10.1111/pbi.13472

Boscaiu Neagu MT, Donat Torres MDP, Llinares Palacios JV, Vicente Meana Ó (2012) Stress-tolerant wild plants: a source of knowledge and biotechnological tools for the genetic improvement of stress tolerance in crop plants. Notulae Botanicae Horti Agrobotanici 40(2):323–327. https://doi.org/10.15835/nbha4028199

Budak H, Kantar M, Yucebilgili Kurtoglu K (2013) Drought tolerance in modern and wild wheat. Sci World J 2013(1):548246. https://doi.org/10.1155/2013/548246

Cabell JF, Oelofse M (2012) An indicator framework for assessing agroecosystem resilience. Ecol Soc 17(1). https://doi.org/10.5751/es-04666-170118

Campos EV, De Oliveira JL, Pascoli M, De Lima R, Fraceto LF (2016) Neem oil and crop protection: from now to the future. Front Plant Sci 7:216834. https://doi.org/10.3389/fpls.2016.01494

Carranza-Gallego G, Guzmán GI, Soto D, Aguilera E, Villa I, Infante-Amate J et al (2018) Modern wheat varieties as a driver of the degradation of Spanish rainfed Mediterranean agroecosystems throughout the 20th century. Sustain For 10(10):3724. https://doi.org/10.3390/su10103724

Colmenares OM, Brindis RC, Verduzco CV, Grajales MP, Gómez MU (2020) Horticultural agroforestry systems recommended for climate change adaptation: a review. Agric Rev 41(1):14–24. https://doi.org/10.18805/ag.r-133

Coyne MS, Pena-Yewtukhiw EM, Grove JH, Sant'Anna AC, Mata-Padrino D (2022) Soil health–it's not all biology. Soil Secur 6:100051–100051. https://doi.org/10.1016/j.soisec.2022.100051

Daily GC, Alexander SE, Ehrlich PR, Goulder L, Lubchenco J, Matson PA, Mooney HA, Postel S, Schneider SH, Tilman D, Woodwell GM (2007) Ecosystem services: benefits supplied to human societies by natural ecosystems. https://www.researchgate.net/profile/Jane_Lubchenco/

publication/200032845_Ecosystem_Services_Benefits_Supplied_to_Human_Societies_by_Natural_Ecosystems/links/0046352cb01672c4b5000000.pdf

Degrune F, Boeraeve F, Dufrêne M, Cornélis JT, Frey B, Hartmann M (2019) The pedological context modulates the response of soil microbial communities to agroecological management. Front Ecol Evol 7:261. https://doi.org/10.3389/fevo.2019.00261

Dighton J, Krumins JA (2014) Interactions in soil: promoting plant growth, 181–202. https://doi.org/10.1007/978-94-017-8890-8

Dotaniya CK, Yashona DS, Aher SB, Rajput PS, Doutaniya RK, Lata M, Mohbe S (2020) Crop performance and soil properties under organic nutrient management. Int J Curr Microbiol App Sci 9(4):1055–1065. https://doi.org/10.20546/ijcmas.2020.904.125

Eisenhauer N, Hines J, Isbell F, van der Plas F, Hobbie SE, Kazanski CE et al (2018) Plant diversity maintains multiple soil functions in future environments. elife 7:e41228. https://doi.org/10.7554/elife.41228

Estrada-Carmona N, Sánchez AC, Remans R, Jones SK (2022) Complex agricultural landscapes host more biodiversity than simple ones: a global meta-analysis. Proc Natl Acad Sci 119(38):e2203385119. https://doi.org/10.1073/pnas.2203385119

Fontaine S, Abbadie L, Aubert M, Barot S, Bloor JM, Derrien D et al (2024) Plant–soil synchrony in nutrient cycles: learning from ecosystems to design sustainable agrosystems. Glob Chang Biol 30(1):e17034. https://doi.org/10.1111/gcb.17034

Fusco G, Melgiovanni M, Porrini D, Ricciardo TM (2020) How to improve the diffusion of climate-smart agriculture: what the literature tells us. Sustain For 12(12):5168. https://doi.org/10.3390/su12125168

Gaba S, Lescourret F, Boudsocq S, Enjalbert J, Hinsinger P, Journet EP, Ozier-Lafontaine H (2015) Multiple cropping systems as drivers for providing multiple ecosystem services: from concepts to design. Agron Sustain Dev 35:607–623. https://doi.org/10.1007/s13593-014-0272-z

Garibaldi LA, Andersson GK, Requier F, Fijen TP, Hipólito J, Kleijn D, Rollin O (2018) Complementarity and synergisms among ecosystem services supporting crop yield. Glob Food Sec 17:38–47. https://doi.org/10.1016/j.gfs.2018.03.006

Gliessman SR (1995) 3 Sustainable agriculture: an agroecological perspective. In: Advances in plant pathology, vol 11. Academic Press, pp 45–57. https://doi.org/10.1016/s0736-4539(06)80005-x

Guo M (2021) Soil health assessment and management: recent development in science and practices. Soil Syst 5(4):61. https://doi.org/10.3390/soilsystems5040061

Gupta J (2010) A history of international climate change policy. Wiley Interdiscip Rev Clim Chang 1(5):636–653. https://doi.org/10.1002/wcc.67

Haddad NM, Brudvig LA, Clobert J, Davies KF, Gonzalez A, Holt RD, Lovejoy TE, Sexton JO, Austin MP, Collins CD, Cook WM, Damschen EI, Ewers RM, Foster BL, Jenkins CN, King AJ, Laurance WF, Levey DJ, Margules C et al (2015, March 6) Habitat fragmentation and its lasting impact on Earth's ecosystems. Am Assoc Adv Sci 1(2):e1500052. https://doi.org/10.1126/sciadv.1500052

He L, Wang S, Peng C, Tan Q (2018) Optimization of water consumption distribution based on crop suitability in the middle reaches of Heihe River. Sustain For 10(7):2119. https://doi.org/10.3390/su10072119

Hellin J, Fisher E (2019) The Achilles heel of climate-smart agriculture. Nat Clim Chang 9(7):493–494. https://doi.org/10.1038/s41558-019-0515-8

Hsiao CJ, Sassenrath G, Rice C, Zeglin L, Hettiarachchi G (2017) Key components of healthy soils and their role in crop production. Kansas Agric Exp Stat Res Reports 3(2):16. https://doi.org/10.4148/2378-5977.1385

Jat SL, Suby SB, Parihar CM, Gambhir G, Kumar N, Rakshit S (2021) Microbiome for sustainable agriculture: a review with special reference to the corn production system. Arch Microbiol 203(6):2771–2793. https://doi.org/10.1007/s00203-021-02320-8

Jose S (2009) Agroforestry for ecosystem services and environmental benefits: an overview. Agrofor Syst 76:1–10. https://doi.org/10.1007/s10457-009-9229-7

Kaushal N, Bhandari K, Siddique KH, Nayyar H (2016) Food crops face rising temperatures: an overview of responses, adaptive mechanisms, and approaches to improve heat tolerance. Cogent Food Agric 2(1):1134380. https://doi.org/10.1080/23311932.2015.1134380

Key G, Whitfield MG, Cooper J, De Vries FT, Collison M, Dedousis T et al (2016) Knowledge needs, available practices, and future challenges in agricultural soils. Soil 2(4):511–521. https://doi.org/10.5194/soil-2-511-2016

Khadka K, Earl HJ, Raizada MN, Navabi A (2020) A physio-morphological trait-based approach for breeding drought tolerant wheat. Front Plant Sci 11:511313. https://doi.org/10.3389/fpls.2020.00715

Kooten OV, Heuvelink E, Stanghellini C (2008) New developments in greenhouse technology can mitigate the water shortage problem of the 21st century XXVII International Horticultural Congress-IHC 2006: international symposium on sustainability through integrated and organic, pp 45–52. https://doi.org/10.17660/actahortic.2008.767.2

Koskey G, Mburu SW, Awino R, Njeru EM, Maingi JM (2021) Potential use of beneficial microorganisms for soil amelioration, phytopathogen biocontrol, and sustainable crop production in smallholder agroecosystems. Front Sustain Food Syst 5:606308. https://doi.org/10.3389/fsufs.2021.606308

Kumar PK, Bellundagi A, Krishna H, Mallikarjuna MG, Thimmappa RK, Rai N et al (2023) Development of bread wheat (*Triticum aestivum* L) variety HD3411 following marker-assisted backcross breeding for drought tolerance. Front Genet 14:1046624. https://doi.org/10.3389/fgene.2023.1046624

Kumari J (2017) Plant genetic resources: effective utilization for sustainability. In: Advancement in crop improvement techniques 3(2). https://doi.org/10.19080/aibm.2017.03.555609

Kumawat A, Yadav D, Samadharmam K, Rashmi I (2020) Soil and water conservation measures for agricultural sustainability. Soil Moist Import 23. https://doi.org/10.5772/intechopen.92895

Lewis J, Rudnick J (2019) The policy enabling environment for climate smart agriculture: a case study of California. Front Sustain Food Syst 3:31. https://doi.org/10.3389/fsufs.2019.00031

Lin BB (2011) Resilience in agriculture through crop diversification: adaptive management for environmental change. Bioscience 61(3):183–193. https://doi.org/10.1525/bio.2011.61.3.4

Liu L, Zhou W, Guan K, Peng B, Xu S, Tang J et al (2024) Knowledge-guided machine learning can improve carbon cycle quantification in agroecosystems. Nat Commun 15(1):357. https://doi.org/10.1038/s41467-023-43860-5

López-Marqués RL, Nørrevang AF, Ache P, Moog M, Visintainer D, Wendt T et al (2020) Prospects for the accelerated improvement of the resilient crop quinoa. J Exp Bot 71(18):5333–5347. https://doi.org/10.1093/jxb/eraa285

Mayer A, Hausfather Z, Jones AD, Silver WL (2018) The potential of agricultural land management to contribute to lower global surface temperatures. Sci Adv 4(8):eaaq0932. https://doi.org/10.1126/sciadv.aaq0932

Mir RR, Rustgi S, Zhang YM, Xu C (2022) Multi-faceted approaches for breeding nutrient-dense, disease-resistant, and climate-resilient crop varieties for food and nutritional security. Heredity 128(6):387–390. https://doi.org/10.1038/s41437-022-00542-0

Morgan JA, Follett RF, Allen LH, Del Grosso S, Derner JD, Dijkstra F et al (2010) Carbon sequestration in agricultural lands of the United States. J Soil Water Conserv 65(1):6A–13A. https://doi.org/10.2489/jswc.65.1.6a

Mubiru DN, Radeny M, Kyazze FB, Zziwa A, Lwasa J, Kinyangi J, Mungai C (2018) Climate trends, risks and coping strategies in smallholder farming systems in Uganda. Clim Risk Manag 22:4–21. https://doi.org/10.1016/j.crm.2018.08.004

Myers N (1996) Environmental services of biodiversity. Proc Natl Acad Sci 93(7):2764–2769. https://doi.org/10.1073/pnas.93.7.2764

Nair PR (2007) The coming of age of agroforestry. J Sci Food Agric 87(9):1613–1619. https://doi.org/10.1002/jsfa.2897

Nair PKR (2013) Agroforestry: trees in support of sustainable agriculture. In: Sustainable agriculture (pp 270–282). Reference module in earth systems and environmental sciences. https://doi.org/10.1016/b978-0-12-409548-9.05088-0
Nair PKR (2014) Agroforestry: practices and systems, an introduction to agroforestry, 270–282. https://doi.org/10.1016/b978-0-444-52512-3.00021-8
Nayak B, Rath BS, Shahid M, Jena SN, Bagchi TB, Roy PS (2020) Organic nutrient management in aromatic rice-rice sequence: a critical review. Int J Chem Stud 8:1435–1444. https://doi.org/10.22271/chemi.2020.v8.i5t.10503
Niu Y, Xiang Y (2018) An overview of biomembrane functions in plant responses to high-temperature stress. Front Plant Sci 9:366067. https://doi.org/10.3389/fpls.2018.00915
Nugroho M, Riniwati H, Afandhi A (2019, February) Model of community empowerment of springs preservation in Arjuna Mountains. In: IOP Conference series: earth and environmental science, vol 239, No 1. IOP Publishing, p 012044. https://doi.org/10.1088/1755-1315/239/1/012044
O'Connor N, Mehta K (2016) Modes of greenhouse water savings. Proc Eng 159:259–266. https://doi.org/10.1016/j.proeng.2016.08.172
Ojo AO, Nwosu NJ, Oshunsanya SO, Ayantayo-Ojo VI, Aladele SE (2023) Impacts of soil conservation techniques on soil erodibility on an Alfisol. Heliyon 9(3). https://doi.org/10.1016/j.heliyon.2023.e13768
Ong C, Black CR, Wilson J, Muthuri C, Bayala J, Jackson NA (2014) Agroforestry: hydrological impacts. https://doi.org/10.1016/b978-0-444-52512-3.00028-0
Oyana TJ, Kayendeke E, Bamutaze Y, Kisanga D (2015) A field assessment of land use systems and soil properties at varied landscape positions in a fragile ecosystem of Mount Elgon, Uganda. African Geogr Rev 34(1):83–103. https://doi.org/10.1080/19376812.2014.929970
Pinho RCD, Miller RP, Alfaia SS (2012) Agroforestry and the improvement of soil fertility: a view from amazonia. Hindawi Publ Corp Appl Environ Sci 2012:1–11. https://doi.org/10.1155/2012/616383
Popp J, Pető K, Nagy J (2013) Pesticide productivity and food security. A review. Agron Sustain Dev 33:243–255. https://doi.org/10.1007/s13593-012-0105-x
Power AG (2010) Ecosystem services and agriculture: tradeoffs and synergies. Philos Trans R Soc B Biol Sci 365(1554):2959–2971. https://doi.org/10.1098/rstb.2010.0143
Raseduzzaman MD, Jensen ES (2017) Does intercropping enhance yield stability in arable crop production? A meta-analysis. Eur J Agron 91:25–33. https://doi.org/10.1016/j.eja.2017.09.009
Rashid S, Mushtaq MB, Farooq I, Khan Z (2021) Climate smart crops for food security. In: The nature, causes, effects and mitigation of climate change on the environment. https://doi.org/10.5772/intechopen.99164
Raza A, Bashir S, Khare T, Karikari B, Copeland RG, Jamla M et al (2024) Temperature-smart plants: a new horizon with omics-driven plant breeding. Physiol Plant 176(1):e14188. https://doi.org/10.1111/ppl.14188
Saura S, Pascual-Hortal L (2007) A new habitat availability index to integrate connectivity in landscape conservation planning: comparison with existing indices and application to a case study. Landsc Urban Plan 83(2–3):91–103. https://doi.org/10.1016/j.landurbplan.2007.03.005
Sethuraman G, Mohd Zain NA, Yusoff S, Ng YM, Baisakh N, Cheng A (2021) Revamping ecosystem services through agroecology—the case of cereals. Agriculture 11(3):204. https://doi.org/10.3390/agriculture11030204
Shukla V, Mattoo AK (2013) Developing robust crop plants for sustaining growth and yield under adverse climatic changes. In: Climate change and plant abiotic stress tolerance, pp 27–56. https://doi.org/10.1002/9783527675265.ch02
Singh RP, Chintagunta AD, Agarwal DK, Kureel RS, Kumar SJ (2020) Varietal replacement rate: prospects and challenges for global food security. Glob Food Sec 25:100324. https://doi.org/10.1016/j.gfs.2019.100324
Sova C, Grosjean G, Baedeker T, Nguyen TN, Wallner M, Nowak A, Corner-Dolloff C, Girvetz E, Läderach P, Lizarazo M (2018) Bringing the concept of climate-smart agriculture to life. In: Impacts, adaptation and vulnerability. https://doi.org/10.1596/31064

Steenwerth KL, Hodson AK, Bloom AJ, Carter MR, Cattaneo A, Chartres C, Hatfield JL, Henry K, Hopmans JW, Horwáth WR, Jenkins BM, Kebreab E, Leemans R, Lipper L, Lubell M, Msangi S, Prabhu R, Reynolds MP, Solís SS, Jackson LE (2014) Climate-smart agriculture global research agenda: Scientific basis for action. Agric Food Secur Agron Sustain Dev 33, 81–96. 3(1), 11. https://doi.org/10.1186/2048-7010-3-11

Stokes A, Douglas GB, Fourcaud T, Giadrossich F, Gillies C, Hubble T et al (2014) Ecological mitigation of hillslope instability: ten key issues facing researchers and practitioners. Plant Soil 377:1–23. https://doi.org/10.1007/s11104-014-2044-6

Stroosnijder L, Moore D, Alharbi A, Argaman E, Biazin B, van den Elsen E (2012) Improving water use efficiency in drylands. Curr Opin Environ Sustain 4(5):497–506. https://doi.org/10.1016/j.cosust.2012.08.011

Swami S (2020) Soil health management under organic production system. IJCS 8(2):330–339. https://doi.org/10.22271/chemi.2020.v8.i2e.8789

Tamburini G, De Simone S, Sigura M, Boscutti F, Marini L (2016) Soil management shapes ecosystem service provision and trade-offs in agricultural landscapes. Proc R Soc B Biol Sci 283(1837):20161369. https://doi.org/10.1098/rspb.2016.1369

Tardieu F, Simonneau T, Muller B (2018) The physiological basis of drought tolerance in crop plants: a scenario-dependent probabilistic approach. Annu Rev Plant Biol 69:733–759. https://doi.org/10.1146/annurev-arplant-042817-040218

Teague WR (2018) Forages and pastures symposium: Cover crops in livestock production: whole-system approach: managing grazing to restore soil health and farm livelihoods. J Anim Sci 96(4):1519–1530. https://doi.org/10.1093/jas/skx060

Velasco L, Ruiz L, Galipienso L, Rubio L, Janssen D (2020) A historical account of viruses in intensive horticultural crops in the Spanish Mediterranean Arc: new challenges for a sustainable agriculture. Agronomy 10(6):860. https://doi.org/10.3390/agronomy10060860

Winter KB, Lincoln NK, Berkes F, Alegado RA, Kurashima N, Frank KL, Pascua P, Rii YM, Reppun F, Knapp ISS, McClatchey WC, Ticktin T, Smith C, Franklin EC, Oleson K, Price MR, McManus MA, Donahue MJ, Rodgers KS, Bowen BW, Nelson C, Thomas B, Leong J-A, Madin EMP, Rivera MAJ, Falinski KA, Bremer LL, Deenik JL, Gon SM III, Neilson B, Okano R, Olegario A, Nyberg B, Kawelo AH, Kotubetey K, Kukea-Shultz JK, Toonen RJ (2020) Ecomimicry in Indigenous resource management: optimizing ecosystem services to achieve resource abundance, with examples from Hawaiʻi. Ecol Soc 25(2):26. https://doi.org/10.5751/ES-11539-250226

Yumurtaç A (2015) Utilization of wild relatives of wheat, barley, maize and oat in developing abiotic and biotic stress tolerant new varieties. Emirates J Food Agric 27(1):1. https://doi.org/10.9755/ejfa.v27i1.17852

Zhan J, Thrall PH, Papaïx J, Xie L, Burdon JJ (2015) Playing on a pathogen's weakness: using evolution to guide sustainable plant disease control strategies. Annu Rev Phytopathol 53:19–43. https://doi.org/10.1146/annurev-phyto-080614-120040

Zhang W, Elias M, Meinzen-Dick R, Swallow K, Calvo-Hernandez C, Nkonya E (2021) Soil health and gender: why and how to identify the linkages. Int J Agric Sustain 19(3–4):269–287. https://doi.org/10.1080/14735903.2021.1906575

Zhao J, Chen J, Beillouin D, Lambers H, Yang Y, Smith P et al (2022) Global systematic review with meta-analysis reveals yield advantage of legume-based rotations and its drivers. Nat Commun 13(1):4926. https://doi.org/10.1038/s41467-022-32464-0

Zomer RJ, Bossio DA, Sommer R, Verchot LV (2017) Global sequestration potential of increased organic carbon in cropland soils. Sci Rep 7(1):15554. https://doi.org/10.1038/s41598-017-15794-8

Chapter 4
The Geography of Agriculture in Sub-Saharan Africa

The agricultural landscape across sub-Saharan Africa has multiple agroecological regions, diverse climate patterns, and a broad spectrum of farming methods. Presented below are fundamental elements outlining the agricultural geography of sub-Saharan Africa:

4.1 Agroecological Diversity

Sub-Saharan Africa encompasses a diverse array of agroecological zones, each with distinct environmental characteristics that shape agricultural practices and crop suitability. The tropical rainforests of the Congo Basin, exemplified by copious rainfall and consistently warm temperatures, facilitate the cultivation of a wide range of crops, including tropical fruits and cash crops like cocoa and oil palm. However, the luxuriant vegetation and delicate ecosystems in these regions present challenges for extensive-scale agriculture, leading to the adoption of traditional farming techniques such as shifting cultivation and agroforestry.

In contrast, the savannas that cover vast areas in countries like Nigeria, Tanzania, and Zambia are characterized by a mix of grassland and scattered trees or shrubs, with distinct wet and dry seasons and lower annual rainfall compared to rain forest regions. Crops commonly grown in these savanna areas include maize, sorghum, millet, and groundnuts, while livestock farming, particularly cattle rearing, is also prevalent due to the availability of grazing lands (Saghir 2014).

Highland areas in sub-Saharan Africa, such as those found in Ethiopia, Kenya, and Lesotho, are characterized by cooler temperatures due to higher elevation, providing more favorable conditions for the cultivation of temperate crops, including potatoes, wheat, barley, and temperate fruits like apples and strawberries (Jayne et al. 2018). Terraced agriculture is commonly practiced in these highland regions

M. Otieno, *Sustainable Agroecological Practices in Sub-Saharan Africa in the Face of Climate Change*, Advances in Geographical and Environmental Sciences, https://doi.org/10.1007/978-3-031-70472-7_4

to optimize land use and mitigate the challenges posed by the rugged terrain (Jayne et al. 2018).

The diversity of agroecological zones in sub-Saharan Africa is further compounded by the region's vulnerability to the impacts of climate change. Studies have shown that climatic changes, such as warmer temperatures and altered rainfall patterns, can have significant impacts on crop productivity and suitability, posing a threat to food security and the livelihoods of smallholder farmers in sub-Saharan Africa (Jayne et al. 2018; Devenish et al. 2023). To address these challenges, building the skills, knowledge, and capacity of rural communities for modernizing agricultural systems, as well as identifying climate-smart strategies, will be crucial for fostering sustainable development and food security in the region (Jayne et al. 2018; Devenish et al. 2023).

The savannas of sub-Saharan Africa, which cover vast areas in countries like Nigeria, Tanzania, and Zambia, face a distinct set of challenges due to the region's vulnerability to climate change. These ecosystems, characterized by a mix of grassland and scattered trees or shrubs, typically experience distinct wet and dry seasons with lower annual rainfall compared to rainforest regions. Crops commonly grown in savanna areas, such as maize, sorghum, millet, and groundnuts, are susceptible to the impacts of climate variability (Jayne et al. 2018).

In addition, the threat of invasive pests and weeds, exacerbated by rising temperatures, further impacts crop production in these areas. Livestock farming, particularly cattle rearing, is also prevalent in savanna regions due to the availability of grazing lands, but this practice may face challenges in the face of changing climatic conditions.

To build resilience and ensure sustainable agricultural development in sub-Saharan Africa's savanna regions, a multifaceted approach is required. This includes adopting climate-smart agricultural practices, improving soil management, and diversifying crop and livestock production systems. Furthermore, building the skills and knowledge of rural communities, as well as strengthening institutional support and access to resources, will be crucial in empowering smallholder farmers to adapt to the evolving climatic challenges.

Overall, the diversity of agroecological zones in sub-Saharan Africa, each with its unique environmental characteristics and agricultural practices, highlights the complex and dynamic nature of the region's food systems. The tropical rainforests, exemplified by the Congo Basin, facilitate the cultivation of a diverse range of crops, including tropical fruits and cash crops, but the delicate ecosystems present challenges for extensive-scale agriculture, leading to the adoption of traditional farming methods. Conversely, the savanna regions, while suitable for the production of staple crops and livestock rearing, face increasing threats from climate change, invasive species, and land degradation. To address these challenges and build resilient and sustainable food systems in sub-Saharan Africa, a multipronged approach is essential, encompassing the adoption of climate-smart agricultural practices, capacity building for rural communities, and strengthening institutional support and access to resources (Jayne et al. 2018; Gugissa et al. 2022).

As the region navigates the complexities of its diverse agroecological zones, the need to develop context-specific and holistic solutions becomes increasingly paramount.

4.2 Addressing the Challenges of Rainfed Agriculture in Sub-Saharan Africa

The majority of agricultural activities in sub-Saharan Africa heavily depend on rainfed farming systems, where rainfall plays a pivotal role in crop growth and development. Rainfed agriculture refers to farming practices that rely on natural precipitation rather than irrigation systems, influenced by limited access to irrigation infrastructure, the predominance of small-scale farming, and the high costs associated with irrigation methods (Jayne et al. 2018). Rainfall patterns in the region exhibit significant variations across different agroecological zones and seasons, with the availability, timing, duration, and intensity of rainfall events greatly influencing agricultural practices and determining planting and harvesting schedules.

Farmers in sub-Saharan Africa typically rely on the onset of the rainy season to prepare their fields and sow crops, as water availability is essential for seed germination and subsequent crop growth. The length of the growing season and the quantity of rainfall received during this period are critical factors that determine optimal yields (Office 2006). However, sub-Saharan Africa is highly vulnerable to climate variability and extreme weather events, including droughts and erratic rainfall patterns (AbdelRahman 2023). Droughts, in particular, pose substantial challenges to agricultural productivity and food security, as insufficient rainfall and prolonged dry spells can lead to crop failures, reduced yields, livestock losses, and water scarcity for both farming and household needs (Calzadilla et al. 2013; Nhemachena et al. 2020; Devenish et al. 2023). The impacts of climate change further exacerbate these issues, with climate projections indicating increased temperatures and changes in precipitation patterns, including more frequent and intense droughts in some areas. These changes pose significant risks to agricultural production and livelihoods as farmers struggle to adapt to the shifting climate conditions (Calzadilla et al. 2013; Devenish et al. 2023).

Efforts to address the challenges of rainfed agriculture in sub-Saharan Africa have focused on a range of strategies, including the promotion of climate-smart agricultural practices, the development of drought-resistant crop varieties, and the improvement of water management and irrigation systems (Calzadilla et al. 2013; Jayne et al. 2018; Devenish et al. 2023). For instance, some studies have examined the potential of input subsidy programs and other policy interventions to encourage the adoption of climate-smart technologies and practices that enhance the resilience of rainfed farming systems. Additionally, researchers have explored the importance of building the skills, knowledge, and capacity of rural communities to modernize their agricultural systems and adapt to the impacts of climate change (Devenish et al. 2023).

4.3 The Role of Traditional Techniques in Subsistence Farming

Subsistence farming remains a fundamental livelihood strategy for the majority of rural households in sub-Saharan Africa, where small-scale farmers engage in predominantly low-input agriculture to meet their own consumption needs. Traditional farming techniques, such as shifting cultivation and mixed cropping, play a significant role in the agricultural landscape of many rural communities in the region.

Shifting cultivation, also known as slash-and-burn agriculture, is a common practice in sub-Saharan Africa. In this method, farmers clear small plots of land, burn the vegetation to release nutrients into the soil, and cultivate crops for a limited period until the soil fertility declines. They then move on to clear another area of land for cultivation. This practice allows for the natural regeneration of the cleared land while farmers cultivate a different plot, enabling sustainable land use. Another traditional farming technique widely employed in the region is mixed cropping, where multiple crops are cultivated together on the same plot of land. This method maximizes land productivity, reduces the risk of crop failure, and provides a diverse range of food sources for subsistence farmers. It also enhances resilience against pests and diseases, as different crops have varying nutrient requirements and growth patterns.

These traditional farming techniques play a crucial role in ensuring food security and sustaining livelihoods in sub-Saharan Africa. Subsistence farming remains a fundamental livelihood strategy for the majority of rural households in the region, where small-scale farmers rely on these low-input agricultural practices to meet their own consumption needs (Asfaw et al. 2012). Shifting cultivation, or slash-and-burn agriculture, is a common practice across various countries in sub-Saharan Africa, including Ghana, Nigeria, and Cameroon. The ability of this technique to allow for natural regeneration of cleared land while farmers cultivate different plots demonstrates its sustainability. Similarly, mixed cropping is widely practiced in rural communities throughout the region, including in countries like Kenya, Tanzania, and Uganda (Baiphethi and Jacobs 2009; Bayu 2020).

The reliance on traditional farming techniques in sub-Saharan Africa is driven by various factors. The continued prevalence of subsistence farming in the region can be attributed to its fundamental role in providing food security and livelihoods for rural households (Asfaw et al. 2012). Furthermore, the small-scale structure of agriculture in sub-Saharan Africa is considered an efficient way to use land and labor resources, given the prevailing market conditions and constraints (Larson et al. 2016). The empirical evidence suggests that the small-scale nature of farming in the region is well-suited to the local context, and that scale-neutral technology innovations can be effectively utilized without the need for significant changes in farm structures. This highlights the central role of small farms in rural development strategies, as they provide a path for reducing poverty and building global food supplies without waiting for broader economic transformations.

The significance of subsistence farming and traditional agricultural techniques in sub-Saharan Africa is further underscored by the region's vulnerability to the impacts of climate change. Traditional farming practices, such as mixed cropping, have been shown to enhance the resilience of rural communities against the challenges posed by climate variability. By diversifying crop production and reducing reliance on external inputs, these techniques help mitigate the risks associated with erratic weather patterns and unpredictable environmental conditions.

In addition, the region continues to face challenges in the conservation and sustainable management of soil and water resources, which are critical to the long-term viability of subsistence farming. Despite the existence of successful local-level initiatives, the broader implementation of soil rehabilitation, water harvesting, and land degradation reduction strategies remains a significant challenge that requires policy attention and investment (Tenywa and Bekunda 2009).

Sub-Saharan Africa has a rich agricultural heritage, with several regions specializing in the production of cash crops for export. These cash crops, such as cocoa, coffee, tea, cotton, and tobacco, play a crucial role in the economies of countries like Côte d'Ivoire, Ghana, Kenya, and Malawi.

The cultivation of cash crops is predominantly carried out on large-scale plantations, a farming system known as plantation agriculture. Côte d'Ivoire, for instance, is renowned for its cocoa production and is one of the largest cocoa exporters globally. The country's economy heavily relies on cocoa, which accounts for a significant portion of its foreign exchange earnings. Similarly, Ghana has a long history of cocoa cultivation and is the second-largest producer of cocoa worldwide. Cocoa is a major cash crop for both small-scale farmers and larger plantations in these countries. In Kenya, tea is a prominent cash crop that contributes significantly to the country's economy (Saghir 2014). The Kenyan tea industry is known for producing high-quality tea, which is exported to various international markets (Saghir 2014). Tea cultivation is primarily carried out on large-scale plantations, with small-scale farmers also involved in the production (Saghir 2014). Malawi is another country in sub-Saharan Africa where cash crops play a vital role in the agricultural sector. Tobacco is one of the major cash crops grown in Malawi and contributes significantly to the country's export earnings. Tobacco farming is predominantly conducted on large-scale plantations and involves both domestic and international tobacco companies (Saghir 2014). The production of cash crops, such as cotton, is also significant in various countries across sub-Saharan Africa.

While the cultivation of cash crops on large-scale plantations has contributed to the economic development of these countries, it has also brought about challenges. The growing activities of cocoa farming have left negative environmental liabilities that are now over stretching the capacity of natural areas in the zone (Merem et al. 2020). Soil degradation and the loss of biodiversity are among the consequences of the heavy use of agrochemicals in these plantations. Efforts to revamp the cocoa sector and ensure sustained production are necessary to address these issues.

Cash crops offer income and employment opportunities to the rural economy and can contribute to food security. Farmers also generate capital for management improvements and innovation, and cash crops accelerate the build-up of institutions that enable further commercialization. However, the growing activities of cash crop cultivation, particularly in plantation agriculture, have also brought about environmental challenges, such as soil degradation and loss of biodiversity due to the heavy use of agrochemicals. These issues need to be addressed to ensure the sustainability of these farming systems.

While the cultivation of cash crops on large-scale plantations has contributed to the economic development of these countries, it has also brought about challenges. The growing activities of cocoa farming have left negative environmental liabilities that are now overstretching the capacity of natural areas in the zone. Soil degradation and the loss of biodiversity are among the consequences of the heavy use of agrochemicals in these plantations. Efforts to revamp the cocoa sector and ensure sustained production are necessary to address these issues (Merem et al. 2020).

Livestock rearing, encompassing cattle, sheep, goats, and poultry, holds significant importance within the agricultural systems of sub-Saharan Africa. Livestock farming plays a crucial role in supporting livelihoods and food production across the region, with pastoralism being a prevalent practice in arid and semi-arid areas (Ibeagha-Awemu et al. 2019; Bonilla-Cedrez et al. 2023). Cattle, as one of the primary livestock species, are widely reared throughout sub-Saharan Africa, serving multiple purposes such as providing meat, milk, hides, and draught power for agricultural activities. Countries like Nigeria, Sudan, and Ethiopia have substantial cattle populations and extensive pastoralist communities. Livestock, particularly cattle, contribute to the resilience of rural households by providing a diversified source of income and nutrition.

Sheep and goats also hold significance in the livestock sector of sub-Saharan Africa. These animals are well-adapted to various ecological conditions and are reared for their meat, milk, and fiber. They are especially important in arid and semi-arid regions where they can thrive on limited vegetation and play a crucial role in supporting the livelihoods of pastoralist communities (Ibeagha-Awemu et al. 2019). Countries like Somalia, Kenya, and Niger have substantial populations of sheep and goats (Ibeagha-Awemu et al. 2019).

Poultry farming, including chickens, ducks, and turkeys, is another integral component of livestock rearing in sub-Saharan Africa. Poultry provides a valuable source of protein through meat and eggs, and it is often practiced on a small scale by rural households. Poultry farming can be highly adaptable and accessible, requiring relatively less land and resources compared to larger livestock species, offering opportunities for income generation and improved nutrition, particularly in densely populated areas (Ibeagha-Awemu et al. 2019; Bonilla-Cedrez et al. 2023).

Pastoralism, characterized by the movement of herds in search of grazing lands, is a traditional livestock production system practiced in arid and semi-arid areas (Ibeagha-Awemu et al. 2019; Bonilla-Cedrez et al. 2023). About 60% of rural households in developing countries are estimated to fully or partly depend on livestock for their livelihoods. Livestock production supports society and generates nearly 40% of global agricultural gross domestic product, with about 1. 3 billion people depending on the sector for their livelihoods (Ibeagha-Awemu et al. 2019). The livestock sector is particularly crucial in sub-Saharan Africa, where it serves as a primary source of revenue, wealth, and status for many rural communities (Ibeagha-Awemu et al. 2019). The diversity of cattle, sheep, goat, pig, and chicken breeds in the region has been shaped by a delicate balance between human and natural selection, as well as environmental adaptation (Ibeagha-Awemu et al. 2019; Bonilla-Cedrez et al. 2023).

Livestock are central to the African society and economy, serving diverse roles such as providing food, income, savings, draft power, and manure for crop production. However, despite these benefits, livestock productivity in the region is less than optimal and unsustainable, failing to match the demands of a growing population. The Food and Agriculture Organization estimates the total African population at 1.3 billion in 2017, with a steadily increasing urban population (Ibeagha-Awemu et al. 2019; Bonilla-Cedrez et al. 2023). Improving the productivity and sustainability of the livestock sector in sub-Saharan Africa is crucial to supporting the livelihoods and food security of millions of people who rely on it (Ibeagha-Awemu et al. 2019; Bonilla-Cedrez et al. 2023). Strategies to enhance livestock production should focus on leveraging available resources, engaging key stakeholders, and adapting to the challenges posed by climate change (Ibeagha-Awemu et al. 2019; Bonilla-Cedrez et al. 2023). Addressing livestock health issues, such as widespread diseases, is critical to improving productivity and sustainability (Ibeagha-Awemu et al. 2019; Kassie and Ali 2019; Bonilla-Cedrez et al. 2023). Investing in research and development to develop more resilient and climate-adapted livestock breeds can also help pastoral and agro-pastoral communities become more resilient to the impacts of climate change.

4.4 Pastoral and Arable Farming in Sub-Saharan Africa

The distribution of pastoral and arable farming in sub-Saharan Africa is largely influenced by factors such as climate, land suitability, and cultural practices. In the arid and semi-arid regions of sub-Saharan Africa, pastoralism has emerged as a prevalent and adaptive livelihood strategy employed by communities to sustain their livelihoods. These areas, characterized by limited rainfall and challenging environmental conditions, are unsuitable for extensive crop cultivation, thus driving communities to rely on the movement of herds in search of grazing lands and water sources as a means of survival.

The Sahel region, stretching across countries like Senegal, Mauritania, Mali, Burkina Faso, Niger, Chad, and Sudan, is a prime example of an area where pastoralism thrives. The low and erratic rainfall in this region makes it well-suited for pastoral activities, as pastoralist communities engage in transhumance, the seasonal movement of herds to access grazing lands and water resources (Scholes 2020). Similarly, the Horn of Africa, encompassing countries like Somalia, Djibouti, and parts of Kenya, is another region where pastoralism is extensively practiced. The arid and semi-arid conditions in this area require mobility and adaptation to sustain livestock, leading pastoralist communities to move their herds across vast distances in search of grazing areas, following the pattern of rainfall and vegetation availability (Scholes 2020).

Certain parts of southern Africa, such as Namibia and Botswana, also have arid and semi-arid regions where pastoralism is a key livelihood strategy. These areas, characterized by low and unpredictable rainfall, support extensive livestock production systems, and pastoralist communities rely on the movement of their herds to access grazing lands and water sources (Machan et al. 2022).

The challenges faced by pastoral communities in sub-Saharan Africa, however, are not limited to environmental factors. Factors such as civil unrest, cattle raiding, and the need for protection have forced some communities into more populated settlements, leading to increased strain on natural resources and the risk of disease outbreaks (Njisane et al. 2020). Furthermore, the effects of climate change, including abrupt changes in weather patterns and unreliable rainfall, are anticipated to have significant negative impacts on the viability of the pastoral livelihood, as they directly affect livestock production, vegetation cover, and natural resource management (Machan et al. 2022). Pastoral and agro-pastoral communities in sub-Saharan Africa are particularly vulnerable to the impacts of climate change, as they rely heavily on the natural resource base for their survival.

In contrast to the arid and semi-arid regions dominated by pastoralism, the more humid and sub-humid areas of sub-Saharan Africa are better suited for arable farming. These regions, characterized by higher and more reliable rainfall, provide favorable conditions for the cultivation of various crops, including staple grains, tubers, and cash crops.

4.5 Arable Farming in Sub-Saharan Africa

Arable farming in sub-Saharan Africa is greatly influenced by various agroecological factors, with its prevalence being higher in areas that possess favorable conditions such as higher rainfall, fertile soils, and suitable temperatures (Nyagumbo et al. 2020). These critical environmental factors play a pivotal role in determining the suitability of a region for successful arable agriculture, ultimately impacting the productivity and sustainability of farming systems.

Rainfall is a crucial determinant of agricultural suitability in sub-Saharan Africa, as areas with higher rainfall tend to support more favorable conditions for arable farming (Jayne et al. 2018). Adequate and well-distributed precipitation ensures sufficient moisture for crop growth, reducing the reliance on irrigation and enhancing the availability of water resources for agricultural purposes (Saghir 2014). Regions with higher rainfall, such as the humid and sub-humid zones of sub-Saharan Africa, often exhibit greater potential for successful arable farming due to their capacity to support diverse crop types and higher yields. Additionally, the length of the rainy season is highly correlated with crop yields, and fluctuations in rainfall patterns can significantly impact food availability at the household level, as less than 1% of the total land use for arable farming is irrigated, making crop production vulnerable to changing rainfall patterns (Musitwa and Keatinge 2002). Declining trends in annual rainfall totals have been observed in many regions of sub-Saharan Africa, including western Africa and southwestern Uganda, further exacerbating the challenges faced by rainfed agriculture.

Soil fertility is another key factor that influences the prevalence of arable farming in sub-Saharan Africa, as fertile soils provide essential nutrients and a suitable physical structure for plant growth. Areas with nutrient-rich soils, such as the volcanic soils found in the Great Lakes region of East Africa, are particularly well-suited for successful agricultural activities and the cultivation of various crops, including coffee, tea, and bananas (Musitwa and Keatinge 2002; Jayne et al. 2018). However, land degradation and depletion of soil nutrients pose major challenges for smallholder farmers, especially in drylands, where more than 24 billion tons of fertile soils are lost in crop-prone land each year (AbdelRahman 2023).

Suitable temperatures also play a significant role in determining the prevalence of arable farming in sub-Saharan Africa, as most crops have specific temperature requirements for optimal growth and development. Areas with temperatures within the suitable range for a particular crop are more likely to support successful cultivation. For example, the favorable temperatures found in certain regions, such as the highlands of East Africa, have contributed to the thriving production of crops like coffee and tea (Jayne et al. 2018).

While these agroecological factors contribute to the suitability and prevalence of arable farming in sub-Saharan Africa, the region's vulnerability to the impacts of climate change poses a significant threat to the existing production systems, infrastructure, and food security. Rising temperatures, altered rainfall patterns, and increased occurrence of extreme weather events can significantly disrupt crop production, leading to decreased yields, changes in the distribution of crops, and the emergence of new pest and disease challenges. For instance, the invasion of fall armyworms and Striga weeds, which thrive in warmer conditions, have become a growing concern for maize-producing farmers in the region. Furthermore, climate change-induced droughts and land degradation can further exacerbate the challenges faced by smallholder farmers, underscoring the need for the development of climate-resilient farming systems and the adoption of sustainable agricultural practices (Gugissa et al. 2022).

Maize plant (corn) attacked by the fall armyworm (*Spodoptera frugiperda*)

4.6 The Influence of Climate, Soil, and Natural Vegetation on Livestock and Crops in Sub-Saharan Africa

The distribution of different animal husbandry types and crops in sub-Saharan Africa is largely influenced by a combination of climatic conditions, soil characteristics, and the region's natural vegetation. In arid and semi-arid regions, such as the Sahel and parts of southern Africa, the climate is a key factor shaping the predominant livelihood strategies.

These areas are characterized by limited and irregular rainfall, as well as the presence of drought-resistant vegetation (Ellis and Galvin 1994). Pastoralism has been a well-adapted and prevalent livelihood in these regions, as pastoralists are able to leverage the natural forage resources provided by the arid and semi-arid ecosystems to sustain their livestock herds (Ellis and Galvin 1994).

The Sahel region, spanning several countries across West and Central Africa, is a prime example of this dynamic. The irregular rainfall patterns and frequent droughts make crop cultivation challenging, but these conditions are suitable for extensive livestock rearing (Machan et al. 2022). Pastoralists in the Sahel have traditionally relied on cattle, sheep, and goats, which are well-adapted to the region's limited and drought-resistant vegetation (Machan et al. 2022). Similarly, parts of southern Africa, including Namibia, Botswana, and South Africa, exhibit arid and semi-arid climates that have historically supported pastoral and nomadic livestock production systems focused on cattle, sheep, and goats (Machan et al. 2022).

The influence of soil characteristics and natural vegetation on the distribution of animal husbandry and crop cultivation in sub-Saharan Africa is also significant. In regions with fertile soils and adequate rainfall, crop cultivation becomes a more viable and prevalent livelihood strategy, often coexisting with livestock production in mixed farming systems (Ibeagha-Awemu et al. 2019). For instance, the humid and sub-humid regions of Central and West Africa, as well as the highlands of East Africa, tend to have more favorable soil conditions and precipitation patterns that support a diverse range of food and cash crops (Kindomihou 2020). These areas are characterized by the cultivation of staple crops like maize, cassava, yams, and various vegetables, often in combination with raising livestock such as cattle, goats, and poultry.

The suitability of these regions for pastoralism is further reinforced by the natural vegetation, which is dominated by drought-tolerant plants like grasses and shrubs that can provide forage for the livestock. Pastoralists in these areas have traditionally practiced nomadic or transhumant pastoralism, moving their herds across vast landscapes in search of water and grazing resources to adapt to the limited rainfall and sparse vegetation (Machan et al. 2022).

In contrast, more humid and tropical regions of sub-Saharan Africa, such as the Congo Basin and parts of West Africa, are characterized by higher rainfall, more fertile soils, and the presence of dense, diverse natural vegetation. These conditions are more conducive to crop cultivation and mixed farming systems, where both livestock and crops are integrated into the livelihood strategies of the local communities (Omotayo et al. 1999; Scholes 2020; Machan et al. 2022).

In the humid zones of Nigeria, for example, nomadic pastoralists, primarily the Fulani, have adapted to the prevalence of tsetse fly and the suitability of the environment for crop production by transitioning to a more sedentary lifestyle, establishing grazing orbits within a limited radius of their settlements and diversifying their livelihoods to include arable crop cultivation (Omotayo et al. 1999). This shift in pastoral practices has been driven by the changing ecology and economy of the region, as the Fulani seek to take advantage of the more favorable climatic and soil conditions for agriculture while still maintaining their livestock production.

The distribution of animal husbandry and crop cultivation in sub-Saharan Africa is thus a complex interaction between climatic factors, soil characteristics, and the natural vegetation of a region. The adaptive strategies of local communities, such as the transition from nomadic to more sedentary pastoralism, also play a significant role in shaping the regional patterns of agricultural and livestock production (Omotayo et al. 1999; Houessou et al. 2019).

References

AbdelRahman MAE (2023) An overview of land degradation, desertification and sustainable land management using GIS and remote sensing applications. Rendiconti Lincei Scienze Fisiche e Naturali 34(3):767–808. https://doi.org/10.1007/s12210-023-01155-3

Asfaw S, Shiferaw B, Simtowe F, Lipper L (2012) Impact of modern agricultural technologies on smallholder welfare: evidence from Tanzania and Ethiopia. Food Policy 37(3):283–295. https://doi.org/10.1016/j.foodpol.2012.02.013

Baiphethi MN, Jacobs PT (2009) The contribution of subsistence farming to food security in South Africa. Agrekon 48(4):459–482. https://doi.org/10.1080/03031853.2009.9523836

Bayu EK (2020) Determinant variables for women's participation in soil and water conservation practices in north western Ethiopia: The case of Shebel Berenta District (Woreda), East Gojjam Zone, Amhara National Regional State. Air Soil Water Res 13:1178622120942199. https://doi.org/10.1177/1178622120942199

Bonilla-Cedrez C, Steward P, Rosenstock TS, Thornton P, Arango J, Kropff M, Ramirez-Villegas J (2023) Priority areas for investment in more sustainable and climate-resilient livestock systems. Nat Sustain 6(10):1279–1286. https://doi.org/10.1038/s41893-023-01161-1

Calzadilla A, Zhu T, Rehdanz K, Tol RS, Ringler C (2013) Economywide impacts of climate change on agriculture in sub-Saharan Africa. Ecol Econ 93:150–165. https://doi.org/10.1016/j.ecolecon.2013.05.006

Devenish AJ, Schmitter P, Jellason NP, Esmail N, Abdi NM, Adanu SK et al (2023) One hundred priority questions for the development of sustainable food systems in sub-Saharan Africa. Land 12(10):1879. https://doi.org/10.22541/au.169419355.58766387/v1

Ellis J, Galvin KA (1994) Climate patterns and land-use practices in the dry zones of Africa. Bioscience 44(5):340–349. https://doi.org/10.2307/1312384

Gugissa DA, Abro Z, Tefera T (2022) Achieving a climate-change resilient farming system through push–pull technology: evidence from maize farming systems in Ethiopia. Sustain For 14(5):2648. https://doi.org/10.3390/su14052648

Houessou SO, Dossa LH, Diogo RV, Houinato M, Buerkert A, Schlecht E (2019) Change and continuity in traditional cattle farming systems of West African Coast countries: a case study from Benin. Agric Syst 168:112–122. https://doi.org/10.1016/j.agsy.2018.11.003

Ibeagha-Awemu EM, Peters SO, Bemji MN, Adeleke MA, Do DN (2019) Leveraging available resources and stakeholder involvement for improved productivity of African livestock in the era of genomic breeding. Front Genet 10:421327. https://doi.org/10.3389/fgene.2019.00357

Jayne TS, Sitko NJ, Mason NM, Skole D (2018) Input subsidy programs and climate smart agriculture: current realities and future potential. Clim Smart Agric 251–273. https://doi.org/10.1007/978-3-319-61194-5_12

Kassie BS, Ali SF (2019) Study on the prevalence and economic significance of bovine fasciolosis at Gondar Elfora Abattoir, Gondar, Ethiopia. J Anim Res 9(5):623–631. https://doi.org/10.30954/2277-940x.05.2019.1

Kindomihou VM (2020) Introductory chapter: milestones in grasses and grassland research. Grasses and Grassland Aspects, 1–30. https://doi.org/10.5772/intechopen.90770

Larson DF, Muraoka R, Otsuka K (2016) On the central role of small farms in African rural development strategies. World Bank Policy Research Working Paper, (7710). https://doi.org/10.1596/1813-9450-7710

Machan SN, Agwata JF, Oguge NO (2022) Environmental factors influencing the sustenance of the camel milk value chain in Isiolo County. Northern Kenya Resourc 11(3):27. https://doi.org/10.3390/resources11030027

Merem CE, Twumasi AY, Wesley J, Olagbegi D, Crisler M, Romorno C et al (2020) Exploring cocoa farm land use in the West African Region. Int J Agric For 10(1):19–39. https://doi.org/10.5923/j.ijaf.20201001.03

Musitwa F, Keatinge JDH (2002) Amending traditional crop management practices to accord with pronounced seasonal rainfall variability in south-central Uganda. East Afr Agric For J 68(1):31–37. https://doi.org/10.4314/eaafj.v68i1.1773

Nhemachena C, Nhamo L, Matchaya G, Nhemachena CR, Muchara B, Karuaihe ST, Mpandeli S (2020) Climate change impacts on water and agriculture sectors in Southern Africa: threats and opportunities for sustainable development. Water 12(10):2673. https://doi.org/10.3390/w12102673

Njisane YZ, Mukumbo FE, Muchenje V (2020) An outlook on livestock welfare conditions in African communities—a review. Asian Australas J Anim Sci 33(6):867. https://doi.org/10.5713/ajas.19.0282

Nyagumbo I, Mupangwa W, Chipindu L, Rusinamhodzi L, Craufurd P (2020) A regional synthesis of seven-year maize yield responses to conservation agriculture technologies in Eastern and Southern Africa. Agric Ecosyst Environ 295:106898–106898. https://doi.org/10.1016/j.agee.2020.106898

Omotayo AM, Adu IF, Aina AB (1999) The evolving sedentary lifestyle among nomadic pastoralists in the humid zone of Nigeria: implications for land-use policy. Int J Sustain Dev World Ecol 6(3):220–228. https://doi.org/10.1080/13504509909470012

Saghir J (2014) Global challenges in agriculture and the World Bank's response in Africa. Food Energy Secur 3(2):61–68. https://doi.org/10.1002/fes3.43

Scholes RJ (2020) The future of semi-arid regions: a weak fabric unravels. Climate 8(3):43. https://doi.org/10.3390/cli8030043

Tenywa MM, Bekunda M (2009) Managing soils in Sub-Saharan Africa: challenges and opportunities for soil and water conservation. J Soil Water Conserv 64(1):44A–48A. https://doi.org/10.2489/jswc.64.1.44a

Chapter 5
Energy-Use Efficiency in Sub-Saharan Africa

5.1 The Fate of Fossil Fuels in Sub-Saharan Africa

The future of fossil fuels in sub-Saharan Africa is shaped by a complex interplay of global energy transitions, regional energy demand, climate change considerations, and the availability of alternative energy sources.

Global energy transitions are driving a shift toward renewable energy sources and the decarbonization of energy systems worldwide (Lincoln 2005). This transition is driven by concerns over climate change and the need to reduce greenhouse gas emissions, resulting in increasing international pressure to reduce reliance on fossil fuels and transition toward cleaner energy alternatives. This global shift has significant implications for sub-Saharan Africa, which must navigate the transition while balancing energy needs and sustainable development goals. The region is home to the world's most rapidly urbanizing cities, with the urban population increasing more than 16-fold between 1950 and 2018. This rapid urban growth is a key driver of energy consumption, with energy use on the continent increasing by an average of 3% between 2000 and 2010 and 2.5% between 2010 and 2018. However, energy consumption in sub-Saharan Africa remains among the lowest in the world, at around 17 GJ per capita per year in 2018, nearly five times below the world average.

Fossil fuels, particularly coal and oil, have historically played a crucial role in meeting the region's growing energy demands, but there is increasing recognition of the need for sustainable and cleaner energy sources to support the region's economic development and improve living standards. Considerations related to climate change are further influencing the discussions surrounding the future of fossil fuels in sub-Saharan Africa, as the region is highly vulnerable to the impacts of climate change, including extreme weather events, droughts, and rising temperatures (Spalding-Fecher et al. 2000; Adewuyi et al. 2020; Piabuo and Puatwoe 2020).

M. Otieno, *Sustainable Agroecological Practices in Sub-Saharan Africa in the Face of Climate Change*, Advances in Geographical and Environmental Sciences, https://doi.org/10.1007/978-3-031-70472-7_5

Despite the challenges, the region is also endowed with a wealth of renewable energy resources, such as solar, wind, and geothermal, which present opportunities for a more sustainable energy future (Piabuo and Puatwoe 2020). Transitioning toward these alternative energy sources can help sub-Saharan Africa reduce its reliance on fossil fuels, mitigate the impacts of climate change, and support its sustainable development goals (Ebhota and Inambao 2016).

A handheld tractor consuming less fossil fuel yet more versatile for a small-scale farmer

5.2 The Renewable Energy Transition in Sub-Saharan Africa

The countries of sub-Saharan Africa are actively pursuing efforts to diversify their energy options and promote the adoption of renewable energy sources, including solar, wind, and hydroelectric power, driven by various factors such as the growing accessibility and declining costs of renewable energy technologies, as well as the numerous benefits associated with renewable energy deployment. The accessibility and affordability of renewable energy technologies have improved significantly in recent years, with the declining costs of solar photovoltaic (PV) panels, wind turbines, and hydroelectric systems making renewable energy more economically viable compared to conventional fossil fuel-based energy sources (Chisika and Yeom 2021). This cost reduction has facilitated the adoption of renewable energy solutions in sub-Saharan Africa, making it a more attractive option for meeting the region's energy needs. Furthermore, governments and international organizations are playing a crucial role in supporting renewable energy projects in sub-Saharan Africa. Governments have implemented policies and regulations that promote renewable energy development, such as feed-in tariffs, renewable portfolio standards, and tax incentives, creating favorable conditions for investment and encouraging private sector participation in the renewable energy sector (Chisika and Yeom 2021). International organizations and development partners also provide support for renewable energy projects through funding, technical assistance, and capacity-building initiatives. For instance, the African Development Bank (AfDB) has established the New Deal on Energy for Africa initiative, which aims to mobilize resources and investments to accelerate the deployment of renewable energy projects across the continent (Adewuyi et al. 2020; Chisika and Yeom 2021). The growing accessibility and affordability of renewable energy technologies, coupled with supportive policy frameworks and external funding, have created a conducive environment for the expansion of renewable energy in sub-Saharan Africa, addressing the region's energy needs and promoting sustainable development (Kwaye et al. 2015; Chisika and Yeom 2021).

The transition toward renewable energy in sub-Saharan Africa holds significant potential to address various socioeconomic challenges facing the region. As the world's most rapidly urbanizing continent, the increasing energy demand driven by urban growth in sub-Saharan Africa can be more effectively met through the deployment of renewable energy solutions, which offer a reliable and sustainable alternative to traditional fossil fuel-based power generation. Furthermore, the wide availability of renewable energy resources, such as abundant solar radiation and wind potential, provides an opportunity for the region to harness these abundant natural resources to meet its energy needs (Karekezi 1994; Waindi and Khalid 2011; Kwaye et al. 2015; Chisika and Yeom 2021).

The transition toward renewable energy in sub-Saharan Africa can contribute to broader development goals, including improving energy access, reducing greenhouse gas emissions, and fostering economic growth. The deployment of renewable energy technologies can help expand energy access to underserved communities, particularly in rural areas, where the decentralized nature of renewable energy solutions can bridge the infrastructure gap and provide clean, affordable, and reliable electricity (Karekezi 1994). By reducing reliance on fossil fuels, the transition to renewable energy can also contribute to climate change mitigation efforts, aligning with the region's commitments to global sustainability goals (Adewuyi et al. 2020). Moreover, the renewable energy sector can serve as a driver of economic development, creating new employment opportunities, attracting foreign investment, and fostering the growth of local industries and manufacturing capabilities.

Wind turbines generating clean energy

5.3 The Role of Renewable Energy in Enhancing Agricultural Productivity in Sub-Saharan Africa

Sub-Saharan Africa faces a significant challenge in terms of energy access, with a large proportion of the population lacking access to modern and reliable energy services (Karekezi 1994). This lack of energy access has profound implications for the region's agricultural sector, hampering productivity, processing, and value chain activities, and limiting the potential for agricultural development (Aliyu et al. 2018; Karekezi 1994). Recognizing the crucial role of electricity in driving agricultural growth and rural development, many countries in the region are prioritizing the expansion of energy access, particularly through renewable energy solutions (Asfaw et al. 2012). On-grid approaches involve connecting agricultural areas to the national electricity grid, providing a reliable and continuous source of power, with governments and utility companies investing in grid infrastructure expansion and electrification initiatives targeting the agricultural sector (Karekezi 1994).

Complementing these on-grid efforts, off-grid renewable energy solutions are also gaining momentum in sub-Saharan Africa. These decentralized systems, such as mini-grids and standalone solar systems, can rapidly provide electricity to rural communities and agricultural activities, particularly in remote areas where grid extension is not economically feasible. These off-grid solutions offer advantages in terms of quick deployment, scalability, and the ability to tailor energy services to specific agricultural needs (Karekezi 1994; Islam et al. 2005).

Various initiatives and programs are being implemented to promote renewable energy access in the agricultural sector of sub-Saharan Africa. For instance, the African Development Bank launched the Technologies for African Agricultural Transformation (TAAT) initiative, which aims to provide access to a range of agricultural technologies, including renewable energy solutions, to improve productivity and value addition across the agricultural value chain (Karekezi 1994). Additionally, efforts are underway to explore the potential of combining different renewable energy sources, such as hydropower and other clean energy options, to develop hybrid systems that can more effectively meet the fluctuating energy demands of the agricultural sector in the region (Waindi and Khalid 2011; Ebhota and Inambao 2016).

The deployment of renewable energy solutions, both on-grid and off-grid, can have significant socioeconomic impacts in sub-Saharan Africa's agricultural communities. Increased access to reliable electricity can enhance irrigation systems, mechanization, post-harvest handling, and agro-processing activities, thereby improving agricultural productivity and value addition (Pahwa 2016). Furthermore, the promotion of regional solar power and renewable energy objectives within the broader frameworks of regional trading blocs, such as the Southern African Development Community and the Economic Community of West African States, can contribute to the sustainable development and regional integration of the agricultural sector (Long and Zhang 2020; Chisika and Yeom 2021).

5.4 The Role of Natural Gas in Powering Agricultural Transformation

Sub-Saharan Africa possesses significant natural gas reserves, and the region has been witnessing ongoing investments in gas exploration, production, and infrastructure development. Natural gas is considered a transition fuel due to its lower carbon emissions compared to coal and oil, making it an attractive option for energy generation and various sectors, including agricultural production (Igbokoyi et al. 2015). The abundant natural gas reserves in sub-Saharan Africa offer potential opportunities for the region's energy sector, with countries such as Nigeria, Mozambique, Tanzania, and Equatorial Guinea actively exploring and developing their gas resources.

These investments aim to harness the economic and environmental benefits associated with natural gas utilization. Natural gas is regarded as a cleaner-burning fuel compared to coal and oil, as it emits lower levels of greenhouse gases and air pollutants when combusted, positioning it as a viable option for reducing carbon emissions and mitigating the impacts of climate change, particularly as a transition fuel during the shift toward renewable energy sources (Igbokoyi et al. 2015).

In the context of agricultural production, natural gas can play a significant role in supporting various activities and processes. Natural gas can be utilized for electricity generation, providing reliable power supply for irrigation systems, agro-processing facilities, and cold storage units, enabling efficient water management, improving post-harvest handling, enhancing food processing capabilities, and extending the shelf life of agricultural produce (Demierre et al. 2015; Mohammed et al. 2020). Moreover, natural gas can also be utilized as a feedstock in the production of fertilizers, a crucial input for agricultural productivity. Fertilizer production requires large amounts of energy, and the availability of affordable and reliable natural gas can support the local production of fertilizers, enhancing agricultural output and food security in the region (Leo 2012; Igbokoyi et al. 2015).

Nigeria, a country endowed with abundant natural gas resources, has been at the forefront of natural gas development and utilization in sub-Saharan Africa. The country's proven natural gas reserves are estimated at 170 trillion cubic feet, making it a prime location for harnessing the economic and environmental benefits associated with this resource (Iwuanyanwu 2007; Akpan 2009). The Nigerian government's Vision 20:2020 initiative aims to position the country among the top 20 economies in the world, with the oil and gas industry, including natural gas, expected to play a crucial role in achieving this goal (Iwuanyanwu 2007; Akpan 2009; Leo 2012).

The natural gas industry in Nigeria has been growing and changing, creating opportunities and challenges for sellers, buyers, and traders. To fully capitalize on the country's natural gas resources, Nigeria needs to develop its downstream industries by setting up more industries that could utilize part of the abundant gas resources, such as in petrochemicals and fertilizers, instead of only exporting the gas as liquefied natural gas (LNG). to developed countries (Akpan 2009; Ani and Ikiensikimama 2020). Developing the downstream industries would create job opportunities for Nigerian citizens, aligning with the government's broader development agenda.

5.5 The Implications of Climate Change Mitigation on Energy Choices in Sub-Saharan Africa

The global focus on climate change mitigation and the imperative to reduce greenhouse gas emissions have significant implications for the energy choices of countries in sub-Saharan Africa. The Paris Agreement, a landmark international treaty aimed at limiting global warming, has galvanized global action and shaped the energy landscape in the region.

In response to the goals of the Paris Agreement and growing environmental consciousness, countries in sub-Saharan Africa are increasingly considering alternative energy options and cleaner technologies. This includes efforts to limit new coal projects, recognizing the high carbon intensity of coal and its contribution to greenhouse gas emissions. Instead, there is a shift toward renewable energy sources such as solar, wind, and hydroelectric power, which are considered more environmentally friendly and sustainable.

The shift toward renewable energy in sub-Saharan Africa is driven by several factors, including the region's significant renewable energy potential and the decreasing costs of renewable technologies. Over 40 African states have installed or are in the process of installing solar power projects or other forms of renewable energy sources (Chisika and Yeom 2021), and the promotion of regional solar power and renewable energy objectives is explored under the broader objectives of regional trading blocs, such as the South African Development Community, the Economic Community for West African States, and the East African Community (Chisika and Yeom 2021). The proliferation of solar power in Africa is attributed to the declining costs of renewable energy components and the creation of a conducive environment for investments in the region.

Furthermore, energy efficiency measures are being promoted to reduce energy consumption and minimize greenhouse gas emissions. Improving energy efficiency in various sectors, including agricultural production, can lead to significant energy savings and lower carbon footprints, contributing to more sustainable and climate-resilient agricultural systems.

5.6 The Viability of Renewable Energy in Sub-Saharan Africa's Agricultural Sector

Sub-Saharan Africa is characterized by diverse economies, and the economic viability of alternative energy sources is a crucial factor in the transition away from fossil fuels in the agricultural sector. Several factors, including the availability and affordability of renewable technologies, energy market dynamics, and investment opportunities, shape the decision-making process.

The availability of renewable energy technologies is a key consideration in the transition away from fossil fuels. Sub-Saharan Africa is blessed with abundant renewable energy resources, including solar, wind, hydroelectric, and biomass energy (Karekezi 1994). The region's geographical and climatic conditions make it suitable for harnessing these renewable sources, and the increasing accessibility and declining costs of renewable technologies contribute to their economic viability and attractiveness as alternatives to fossil fuels.

Energy market dynamics also play a significant role in the transition. The evolving energy landscape, changing energy policies, and market incentives influence the economic feasibility of alternative energy sources. Governments and regulatory bodies are implementing measures to promote renewable energy adoption, such as feed-in tariffs, renewable energy targets, and subsidies, which can create a favorable environment for investment in renewable energy projects (Karekezi 1994).

Additionally, investment opportunities and financial mechanisms are critical in driving the transition to alternative energy sources. Public and private investments, including domestic and international financing, are essential for the development and deployment of renewable energy infrastructure. International organizations, development banks, and private investors are increasingly recognizing the economic potential of renewable energy in sub-Saharan Africa and are providing financial support and technical expertise to catalyze the energy transition in the region's agricultural sector (Karekezi 1994; Dauda and Hui 2011; Kwaye et al. 2015; Chisika and Yeom 2021; Amir and Khan 2022).

The energy transition in sub-Saharan Africa's agricultural sector faces several challenges, including issues with power grid restructuring, energy storage technologies, and the mitigation of environmental factors with seasonal variations. However, the region's abundant renewable energy resources, the declining costs of renewable technologies, and the evolving policy and investment landscape suggest that the economic viability of alternative energy sources is a promising opportunity for the region's agricultural sector.

5.7 Alternative Energy Sources

Sub-Saharan Africa has abundant renewable energy resources, and there is growing interest in developing alternative energy sources to meet the region's energy needs.

5.8 The Promising Role of Solar Energy in Sub-Saharan African Agriculture

The use of solar energy for agricultural production in sub-Saharan Africa is a highly promising and rapidly growing practice, primarily due to the region's abundant sunshine and the numerous benefits offered by solar power. Solar energy is being harnessed through various approaches, including large-scale solar farms and decentralized solar systems, to cater to the diverse needs of agricultural activities (Chisika and Yeom 2021; Cisse et al. 2023).

Solar energy plays a vital role in electricity generation for agricultural operations. Solar photovoltaic (PV) systems are installed to power agricultural machinery, equipment, and irrigation systems, helping to reduce reliance on traditional grid electricity and providing a more sustainable and cost-effective energy solution (Li-xue 2017). Water pumping is another crucial application of solar energy in agriculture, with solar-powered irrigation systems being widely utilized to extract water for irrigation purposes, particularly in off-grid or remote areas where access to conventional electricity or diesel fuel is limited (Li-xue 2017; Chisika and Yeom 2021). Solar-powered irrigation helps enhance crop productivity, reduce labor-intensive manual irrigation, and improve water management efficiency. Solar energy is also utilized for lighting and other energy needs in agricultural settings, powering farm buildings, sheds, and facilities, and enabling extended working hours, increased safety, and improved productivity. Large-scale solar farms, consisting of ground-mounted solar panels, also contribute to the renewable energy landscape in sub-Saharan Africa, providing a decentralized and sustainable source of electricity for agricultural activities (Hernandez et al. 2015; Abu-Nowar 2020).

The promotion of regional solar power and renewable energy objectives in Africa is being explored under the broader objectives of regional trading blocs, such as the South African Development Community, the Economic Community for West African States, and the East African Community. The proliferation of solar energy in Africa is attributed to the declining cost of renewable energy, with larger markets like China, Europe, and India closing in on their production targets, making solar components more readily available and creating a conducive environment for investments in regions like sub-Saharan Africa.

A solar panel powering a permaculture farm

5.9 The Potential of Wind Energy in Agricultural Production Across Sub-Saharan Africa

Wind power holds significant potential for agricultural production in specific regions of sub-Saharan Africa, especially coastal areas and high-wind zones. Both large-scale wind farms and small-scale wind turbines are being deployed to harness wind energy for electricity generation, with notable progress observed in countries such as South Africa, Kenya, and Ethiopia.

Large-scale wind farms are being established in coastal regions and other areas with favorable wind conditions. These farms consist of multiple wind turbines strategically positioned to capture the maximum wind energy (Silalahi and Blakers 2023; Wimhurst et al. 2023). They contribute to the overall electricity supply, including for agricultural activities, by feeding clean and sustainable energy into the grid (Kwaye et al. 2015). South Africa, in particular, has made significant investments in large-scale wind power projects, such as the Renewable Energy Independent Power Producer Procurement Program (REIPPPP), which has facilitated the development of numerous wind farms.

Small-scale wind turbines are also gaining prominence in agricultural settings. These turbines, typically installed on farms or rural properties, provide decentralized and off-grid electricity solutions for agricultural operations. They are particularly suitable for remote or isolated areas where grid connectivity is limited. Small-scale wind turbines can power various agricultural applications, including water pumping, lighting, and running small-scale machinery. Kenya and Ethiopia have shown notable progress in the deployment of small-scale wind turbines, often used to support irrigation systems and enhance energy access for rural farming communities.

Numerous studies and reports emphasize the potential and advantages of wind power in agricultural production across sub-Saharan Africa. They highlight the favorable wind resources in coastal and high-wind areas, the technical feasibility of wind power systems, and the potential for these systems to support agricultural activities and improve energy access for rural communities (Karekezi 2002; Kwaye et al. 2015; Boureguig et al. 2020; Wimhurst et al. 2023). The growing adoption of wind power, both at large and small scales, underscores the increasing recognition of its value in enhancing agricultural productivity and sustainability in the region (Lewis and Wiser 2007; Abraham et al. 2012; Boureguig et al. 2020; Wimhurst et al. 2023).

5.10 Harnessing the Power of Hydroelectricity to Agricultural Transformation in Sub-Saharan Africa

Sub-Saharan Africa possesses an abundance of rivers and water bodies that offer immense potential for hydropower generation, which can be harnessed to power various agricultural operations across the region. The development of both large-scale hydroelectric projects and small/mini-hydro schemes is underway, contributing to the region's electricity supply and enabling the utilization of renewable energy sources for agricultural activities.

Large-scale hydroelectric projects, such as the Grand Ethiopian Renaissance Dam in Ethiopia, the Inga Dam complex in the Democratic Republic of Congo, and the Cahora Bassa Dam in Mozambique, are being established on major rivers to harness the substantial hydropower potential of the region (Adu et al. 2018). These projects involve the construction of large dams and power plants that generate electricity from the flow of water, providing a reliable and renewable energy source for agricultural operations (Adu et al. 2018; Kichonge 2018; Yevalla et al. 2018; Pappis et al. 2022).

In addition to these large-scale initiatives, small and mini-hydro schemes are gaining attention for their suitability in decentralized and off-grid settings, particularly in rural areas. These smaller-scale hydroelectric systems, typically installed on rivers or streams with moderate flow rates, can provide localized electricity generation for agricultural purposes, such as powering irrigation systems, machinery, and rural electrification initiatives (Adu et al. 2018; Yevalla et al. 2018).

The prospects and advantages of utilizing hydropower for agriculture in sub-Saharan Africa are widely recognized in various studies and reports. The region's abundant water resources, the technical feasibility of hydropower projects, and the socioeconomic benefits associated with increased energy access and reduced reliance on fossil fuels have been highlighted (Ebhota and Inambao 2016; Adu et al. 2018; Kichonge 2018; Yevalla et al. 2018). Moreover, initiatives by governments, development partners, and research institutions are underway to further harness the potential of hydropower and support its integration into the agricultural sector (Ebhota and Inambao 2016; Adu et al. 2018).

5.11 The Potential of Biomass and Bioenergy in Sub-Saharan Africa's Agricultural Landscapes

Biomass resources in sub-Saharan Africa, including agricultural residues, forestry waste, and organic matter, hold significant potential for utilization in agricultural production. These biomass resources can be harnessed to produce bioenergy, such as biogas and biofuels, which offer alternative energy sources for agroecological systems (Ejigu 2008; Owusu and Asumadu-Sarkodie 2016; Jekayinfa et al. 2020). Agricultural residues, such as crop stalks, husks, and straw, are abundant in agricultural landscapes and can be collected and processed to generate energy in the form of biogas or solid biomass fuels (Ejigu 2008; Owusu and Asumadu-Sarkodie 2016). Biogas production involves the anaerobic digestion of organic matter, such as livestock manure and crop residues, to produce a renewable gas that can be used for cooking, heating, and generating electricity (Ejigu 2008). Biogas digesters are being increasingly adopted in sub-Saharan Africa, particularly in rural areas, to provide clean and affordable energy for cooking and other household needs (Ejigu 2008). Forestry waste, such as branches, sawdust, and wood chips, can also be utilized as biomass resources for energy production. These waste materials can be processed into solid biomass fuels, such as wood pellets or briquettes, which can be used for cooking, heating, and powering agricultural machinery. The use of biomass fuels derived from forestry waste can contribute to reducing reliance on traditional biomass, such as firewood and charcoal, which can have negative environmental and health impacts. Furthermore, organic matter, including animal manure and crop residues, can be converted into biofuels. Biofuels derived from agricultural feedstocks, such as ethanol from sugarcane or biodiesel from oil crops like Jatropha, can be used as a renewable substitute for fossil fuels in agricultural machinery and transportation. Biofuel production from agricultural resources presents an opportunity to promote sustainable agriculture and reduce greenhouse gas emissions. The growing global interest in biofuels has prompted many African countries to start harnessing this form of energy, and it is envisaged that the biofuel sector will be a thriving one in most African countries in the next two decades (Ejigu 2008; Babajide et al. 2015).

5.12 Harnessing Geothermal Potential: Powering Sustainable Agriculture in Sub-Saharan Africa

Sub-Saharan Africa, particularly the East African Rift System, possesses significant geothermal potential, making it a promising renewable energy source for the region. Geothermal power plants are being developed to harness this potential and provide reliable baseload electricity for various sectors, including agricultural production and agroecology (Babatunde et al. 2019; Chisika and Yeom 2021; Pavelic et al. 2021). The East African Rift System, characterized by tectonic activity and volcanic formations, offers favorable conditions for geothermal energy extraction (Babatunde et al. 2019). The region has a substantial number of geothermal reservoirs with high-temperature resources suitable for electricity generation (Moore and Simmons 2013). Countries such as Kenya, Ethiopia, and Tanzania have made notable progress in developing geothermal projects, aiming to diversify their energy mix and reduce reliance on fossil fuels.

Geothermal power plants utilize the heat stored within the Earth's crust to generate electricity. Steam or hot water from geothermal reservoirs is used to drive turbines, producing clean and renewable energy. Geothermal energy offers several advantages for agricultural production and agroecology. It provides a stable and continuous source of electricity, which is crucial for powering irrigation systems, water pumping, and other agricultural processes. The availability of reliable energy facilitates efficient irrigation practices, enhancing crop yields and supporting sustainable agriculture. Geothermal energy is also environmentally friendly, emitting minimal greenhouse gases compared to fossil fuel-based power generation. By utilizing geothermal power for agricultural production, the sector can reduce its carbon footprint and contribute to climate change mitigation efforts (Moore and Simmons 2013; Babatunde et al. 2019).

While the geothermal potential in sub-Saharan Africa is significant, there are still challenges to overcome, such as the difficulty in obtaining accurate data in undeveloped and developing nations, high costs of initial investments, lack of technical skills, and lack of societal awareness and incentives. Nevertheless, the region's commitment to harnessing renewable energy sources, including geothermal, holds promise for a more sustainable and resilient agricultural sector.

5.13 Enhancing Energy Efficiency in Agriculture: Strategies for Reducing Inputs Without Compromising Outputs

Reducing energy inputs without affecting energy outputs is a crucial objective for improving the sustainability and efficiency of agricultural systems. This can be achieved through the implementation of various energy efficiency measures and optimization of agricultural processes.

One important strategy is the upgrading of agricultural equipment and infrastructure. This includes replacing outdated or inefficient machinery with more energy-efficient models, such as transitioning to energy-efficient irrigation systems like drip or precision irrigation, which can significantly reduce water and energy losses (Bakker et al. 2008). Similarly, enhancing the insulation and design of agricultural buildings and greenhouses can minimize heat loss or gain, lowering the energy demands for heating and cooling (Bakker et al. 2008).

The optimization of agricultural processes is another key approach. This involves streamlining workflows, minimizing energy-intensive operations, and adopting precision farming techniques that utilize advanced technologies like remote sensing and geographic information systems (Nhemachena et al. 2020). By optimizing processes, farmers can achieve greater energy efficiency without compromising productivity (Bhattacharya et al. 2020; Nhemachena et al. 2020).

Synchronizing nutrient cycles within agricultural systems can also contribute to energy efficiency. Aligning nutrient availability with plant uptake can reduce reliance on external inputs and minimize energy-intensive processes like fertilizer production and application (Fontaine et al. 2024). Furthermore, climate-smart agriculture practices that enhance resilience and adaptive capacity can help mitigate the impacts of climate change on water and energy resources in vulnerable regions (Shah and Wu 2019; Bhattacharya et al. 2020; Nhemachena et al. 2020; Fontaine et al. 2024).

Ultimately, the path to enhancing energy efficiency in agriculture requires a multifaceted approach that combines technological innovations, optimized processes, and nature-based solutions. By adopting these strategies, farmers and agro-industries can reduce energy inputs while maintaining or even improving energy outputs, leading to more sustainable and resilient agricultural systems (Pretty 2008; Scialabba and Müller-Lindenlauf 2010; Bhattacharya et al. 2020; Shah and Wu 2019; Nhemachena et al. 2020; Fontaine et al. 2024).

5.14 Smart Energy Management Systems in Agriculture

Utilizing advanced monitoring and control systems in agricultural systems is an effective approach to optimize energy usage and minimize wastage. These systems leverage smart agricultural tools, energy management systems, and building automation technologies to enhance energy efficiency and productivity in agricultural operations. By incorporating real-time monitoring, data analysis, and automation, energy consumption can be regulated based on demand, standby power can be minimized, and load shifting can be enabled to achieve maximum efficiency.

Smart agricultural tools, such as sensor networks and Internet of Things (IoT) devices, provide valuable data on environmental conditions, crop health, and energy usage. By collecting and analyzing this data, farmers can make informed decisions to optimize energy usage. For instance, soil moisture sensors can provide information about the water content in the soil, enabling precise irrigation scheduling and avoiding unnecessary energy expenditure on over-irrigation. Similarly, weather monitoring stations can provide accurate weather forecasts, allowing farmers to adjust energy-intensive operations like spraying or harvesting based on upcoming weather conditions, thereby reducing energy waste.

Energy management systems play a vital role in regulating and optimizing energy consumption in agricultural systems. These systems enable real-time monitoring of energy usage, identify energy-intensive processes, and provide insights for energy-saving opportunities. By integrating energy management systems with smart meters and automated control systems, farmers can monitor energy consumption, identify energy-saving potential, and implement demand-response strategies to shift energy usage to off-peak hours. This load shifting approach helps reduce peak energy demand and promotes efficient use of energy resources. Building automation technologies, such as smart irrigation systems and automated lighting controls, further enhance energy efficiency by optimizing resource utilization based on real-time data and changing environmental conditions.

The integration of smart agricultural tools, energy management systems, and building automation technologies can significantly improve energy efficiency and productivity in agricultural operations. By leveraging advanced monitoring, data analysis, and automation, farmers can make informed decisions, optimize energy usage, and reduce energy waste, contributing to more sustainable and cost-effective agricultural practices.

5.15 The Role of Behavior Changes and Awareness in Reducing Energy Inputs in Agriculture

Promoting energy conservation practices in agricultural production and agroecology is crucial for reducing energy inputs without compromising energy outputs. Educational campaigns and programs focused on energy conservation and efficiency have been implemented in various agricultural contexts, aiming to raise awareness about the environmental and economic benefits of energy conservation practices and encourage behavioral changes among farmers and agricultural stakeholders.

Studies have shown that these interventions can have a positive impact on energy conservation practices in agriculture (Liu 2022). For instance, an awareness-raising campaign in Switzerland significantly influenced farmers' energy-saving behavior, leading to reduced energy consumption in dairy farming (Liu 2022). Similarly, educational programs targeting farmers in New Zealand resulted in improved energy efficiency and reduced energy consumption in agricultural buildings.

Adopting simple energy-saving habits, such as turning off unused equipment, using energy-efficient lighting, and proper equipment maintenance, can also contribute to long-term reductions in energy consumption in agriculture (Scialabba and Müller-Lindenlauf 2010; Babatunde et al. 2019). Renewable energy sources can also play a role in curbing emissions and energy use in the agricultural sector. Business-friendly legislation and innovative public benefits funds to encourage the adoption of renewable electricity generation in agriculture can further support these efforts.

References

Abraham J, Plourde BD, Mowry G, Minkowycz W, Sparrow EM (2012) Summary of Savonius wind turbine development and future applications for small-scale power generation. J Renew Sustain Energ 4:042703. https://doi.org/10.1063/1.4747822

Abu-Nowar LM (2020) Economic and financial assessment of solar-powered irrigation. J Agric Sci 12(4):185–185. https://doi.org/10.5539/jas.v12n4p185

Adewuyi OB, Kiptoo MK, Afolayan AF, Amara T, Alawode OI, Senjyu T (2020) Challenges and prospects of Nigeria's sustainable energy transition with lessons from other countries' experiences. Energy Rep 6:993–1009. https://doi.org/10.1016/j.egyr.2020.04.022

Adu D, Zhang J, Fang Y, Jia J, Chen S, Suoming L (2018) The prospects & submissions for small hydropower development in sub-Saharan African. IOP Conf Ser Earth Environ Sci 163:012022. https://doi.org/10.1088/1755-1315/163/1/012022

Akpan S (2009) Production and utilization of natural gas resources in Nigeria – a review. Paper presented at the Nigeria Annual International Conference and Exhibition, Abuja, Nigeria, August 2009. https://doi.org/10.2118/128356-ms

Aliyu AK, Modu B, Tan CW (2018) A review of renewable energy development in Africa: a focus in South Africa, Egypt and Nigeria. Renew Sust Energ Rev 81:2502–2518. https://doi.org/10.1016/j.rser.2017.06.055

Amir M, Khan SZ (2022) Assessment of renewable energy: status, challenges, COVID-19 impacts, opportunities, and sustainable energy solutions in Africa. Energ Built Environ 3(3):348–362. https://doi.org/10.1016/j.enbenv.2021.03.002

Ani GO, Ikiensikimama SS (2020) Natural gas development trend in Nigeria: the investor's road map. J Energ Res Rev 5(4):39–48. https://doi.org/10.9734/jenrr/2020/v5i430155

Asfaw S, Shiferaw B, Simtowe F, Lipper L (2012) Impact of modern agricultural technologies on smallholder welfare: evidence from Tanzania and Ethiopia. Food Policy 37(3):283–295. https://doi.org/10.1016/j.foodpol.2012.02.013

Babajide O, Petrik L, Ameer F (2015) Technologies for biodiesel production in sub-Saharan African countries. In: Biofuels – status and perspective. https://doi.org/10.5772/59859

Babatunde OM, Denwigwe IH, Adedoja OS, Babatunde DE, Gbadamosi SL (2019) Harnessing renewable energy for sustainable agricultural applications. Int J Energy Econ Policy 9(5):308–315. https://doi.org/10.32479/ijeep.7775

Bakker T, Wouters H, van Asselt K, Bontsema J, Tang L, Müller J, van Straten G (2008) A vision based row detection system for sugar beet. Comput Electr Agric (CEA) 60(1):87–95. https://doi.org/10.1002/rob.21995

Bhattacharya U, Sarkar S, Bandopadhyay P (2020) Optimum use of nonmonetary inputs for sustainability in agriculture. Int J Curr Microbiol App Sci 9(4):1185–1195. https://doi.org/10.20546/ijcmas.2020.904.140

Boureguig K, Mansouri A, Chouya A (2020) Performance enhancements of DFIG wind turbine using fuzzy-feedback linearization controller augmented by high-gain observer. Int J Pow Elec Dri Syst ISSN 2088(8694):8694. https://doi.org/10.11591/ijpeds.v11.i1.pp10-23

Chisika SN, Yeom C (2021) Enhancing sustainable development and regional integration through electrification by solar power: the case of six east African states. Sustain For 13(6):3275. https://doi.org/10.3390/su13063275

Cisse EHI, Ndiogou BA, Tigampo S, Thiam A, Azilinon D (2023) Optimization of a solar chimney with a horizontal absorber for building ventilation: a case study. Fluid Dyn Mater Process 19(4):901–910. https://doi.org/10.32604/fdmp.2022.021973

Dauda AM, Hui G (2011, August) Technology transfer for renewable energy development and use in Africa. In: 2011 2nd International Conference on Artificial Intelligence, Management Science and Electronic Commerce (AIMSEC). IEEE, pp 4800–4804. https://doi.org/10.1109/aimsec.2011.6010184

Demierre J, Bazilian M, Carbajal J, Sherpa S, Modi V (2015) Potential for regional use of East Africa's natural gas. Appl Energy 143:414–436. https://doi.org/10.1016/j.apenergy.2015.01.012

Ebhota WS, Inambao FL (2016) Electricity insufficiency in Africa: a product of inadequate manufacturing capacity. Afr J Sci Technol Innov Dev 8(2):197–204. https://doi.org/10.1080/20421338.2016.1147206

Ejigu M (2008, May) Toward energy and livelihoods security in Africa: smallholder production and processing of bioenergy as a strategy. In: Natural resources forum (Vol 32, No 2, pp 152–162). Oxford: Blackwell https://doi.org/10.1111/j.1477-8947.2008.00189.x

Fontaine S, Abbadie L, Aubert M, Barot S, Bloor JM, Derrien D et al (2024) Plant–soil synchrony in nutrient cycles: learning from ecosystems to design sustainable agrosystems. Glob Chang Biol 30(1):e17034. https://doi.org/10.1111/gcb.17034

Hernandez RR, Hoffacker MK, Murphy-Mariscal ML, Wu GC, Allen MF (2015) Solar energy development impacts on land cover change and protected areas. Proc Natl Acad Sci 112(44):13579–13584. https://doi.org/10.1073/pnas.1517656112

Igbokoyi AO, Nigeria A, Iledare O (2015, August) The role of natural gas in Nigeria power generation: empirical analysis of challenges, constraints and opportunities. In SPE Nigeria annual international conference and exhibition (pp SPE-178394). SPE. https://doi.org/10.2118/178394-MS

Islam M, Islam AS, Amin MR (2005, January) Small-scale decentralized renewable energy systems for the remote communities of the developing countries. In: ASME power conference (Vol 41820, pp 879–890). https://doi.org/10.1115/pwr2005-50068

Iwuanyanwu HU (2007, December) Challenges of natural gas utilization in Nigeria. In: International petroleum technology conference (pp IPTC-11177). IPTC. https://doi.org/10.2523/iptc-11177-abstract

Jekayinfa SO, Orisaleye JI, Pecenka R (2020) An assessment of potential resources for biomass energy in Nigeria. Resources 9(8):92. https://doi.org/10.3390/resources9080092

Karekezi S (1994) Disseminating renewable energy technologies in sub-Saharan Africa. Annu Rev Energy Environ 19(1):387–421. https://doi.org/10.1146/annurev.eg.19.110194.002131

Karekezi S (2002) Renewables in Africa—meeting the energy needs of the poor. Energy Policy 30(11–12):1059–1069. https://doi.org/10.1016/s0301-4215(02)00058-7

Kichonge B (2018) The status and future prospects of hydropower for sustainable water and energy development in Tanzania. J Renew Energy 2018(1):6570358. https://doi.org/10.1155/2018/6570358

Kwaye MP, Bendfeld J, Anglani N (2015, May) Assessment of renewable energy resources and the use of hydro power for fluctuation compensation in Cameroon. In: 2015 5th International Youth Conference on Energy (IYCE). IEEE, pp 1–5. https://doi.org/10.1109/iyce.2015.7180806

Leo M (2012, August). Vision 20: 2020 initiative and the Nigerian oil and gas industry: an empirical analysis of the environmental dimension. In SPE Nigeria annual international conference and exhibition (pp SPE-163025). SPE. https://doi.org/10.2118/163025-ms

Lewis JI, Wiser RH (2007) Fostering a renewable energy technology industry: an international comparison of wind industry policy support mechanisms. Energy Policy 35(3):1844–1857. https://doi.org/10.1016/j.enpol.2006.06.005

Lincoln SF (2005) Fossil fuels in the 21st century. Ambio 34:621–627. https://doi.org/10.1579/0044-7447-34.8.621

Liu H (2022, March) Exerting behavioral economics to household energy use. In: 2022 7th International Conference on Financial Innovation and Economic Development (ICFIED 2022). Atlantis Press, pp 1265–1270. https://doi.org/10.2991/aebmr.k.220307.209

Li-xue W (2017) Solar power pumping in agriculture: a review of recent research. Agric Res Technol Open Access J 4(3). https://doi.org/10.19080/artoaj.2017.04.555636

Long C, Zhang X (2020, October) Renewable energy in Africa: a discussion of the feasibility and socio-economic effects of solar photovoltaic (PV) system. In: IOP conference series: earth and environmental science (Vol 586, No 1, 012012). IOP Publishing. https://doi.org/10.1088/1755-1315/586/1/012012

Mohammed YS, Maruf AA, Amlabu CA, Bashayi JG, Sanni YY, Audu N (2020) Estimation of electricity production from corn residues: a case study of Nigeria. ARPN J Eng Appl Sci 15(6):1385–1392. https://doi.org/10.36478/jeasci.2020.1385.1392

Moore JN, Simmons SW (2013) More power from below. Am Assoc Adv Sci 340(6135):933–934. https://doi.org/10.1126/science.1235640

Nhemachena C, Nhamo L, Matchaya G, Nhemachena CR, Muchara B, Karuaihe ST, Mpandeli S (2020) Climate change impacts on water and agriculture sectors in Southern Africa: threats and opportunities for sustainable development. Water 12(10):2673. https://doi.org/10.3390/w12102673

Owusu PA, Asumadu-Sarkodie S (2016) A review of renewable energy sources, sustainability issues and climate change mitigation. Cogent Eng 3(1):1167990. https://doi.org/10.1080/23311916.2016.1167990

Pahwa A (2016, June) Partnerships to facilitate electricity access for the remote rural communities of sub-Sahara Africa. In: 2016 IEEE PES PowerAfrica. IEEE, pp 83–87. https://doi.org/10.1109/powerafrica.2016.7556576

Pappis I, Sridharan V, Howells M, Medarac H, Kougias I, Sánchez RG et al (2022) The effects of climate change mitigation strategies on the energy system of Africa and its associated water footprint. Environ Res Lett 17(4):044048. https://doi.org/10.1088/1748-9326/ac5ede

Pavelic P, Magombeyi M, Schmitter P, Jacobs-Mata I (2021) Sustainable expansion of groundwater-based solar water pumping for smallholder farmers in Sub-Saharan Africa. Washington, DC, USA: Efficiency for Access Coalition. 54p. https://cgspace.cgiar.org/handle/10568/114832

Piabuo SM, Puatwoe JT (2020) Public health effects of wood fuel in Africa: bioenergy from tree commodities as a sustainable remedy. In: Public health in developing countries-challenges and opportunities. https://doi.org/10.5772/intechopen.90603

Pretty J (2008) Agricultural sustainability: concepts, principles and evidence. Philos Trans R Soc B Biol Sci 363(1491):447–465. https://doi.org/10.1098/rstb.2007.2163

Scialabba NEH, Müller-Lindenlauf M (2010) Organic agriculture and climate change. Renew Agric Food Syst 25(2):158–169. https://doi.org/10.1017/s1742170510000116

Shah F, Wu W (2019) Soil and crop management strategies to ensure higher crop productivity within sustainable environments. Sustain For 11(5):1485. https://doi.org/10.3390/su11051485

Silalahi DF, Blakers A (2023, July) Global atlas of marine floating solar PV potential. In: Solar (Vol 3, No 3, pp 416–433). MDPI. https://doi.org/10.3390/solar3030023

Spalding-Fecher R, Williams A, van Horen C (2000) Energy and environment in South Africa: charting a course to sustainability. Energy Sustain Dev 4(4):8–17. https://doi.org/10.1016/s0973-0826(08)60259-8

Waindi IO, Khalid A (2011) Renewable energy in East Africa: an introductory evaluation using a systems approach to assess alternatives to providing electricity. J Afr Bus 12(3):387–418. https://doi.org/10.1080/15228916.2011.621854

Wimhurst JJ, Nsude CC, Greene JS (2023) Standardizing the factors used in wind farm site suitability models: a review. Heliyon. https://doi.org/10.1016/j.heliyon.2023.e15903

Yevalla GMS, Seidou DN, Ngoc BV, Hoi TV, Tabod TC (2018) Hydrological studies for the assessment of run-of-river hydropower potential and generation over the Wouri-Nkam River using GIS and remote sensing techniques. Aquademia Water Environ Technol 2(1):02. https://doi.org/10.20897/awet/86712

Chapter 6
Traditional and Industrial Farming Practices

Traditional farming practices can play a significant role in the principles and practices of agroecology. Traditional farming practices often align with the principles of agroecology due to their long-standing reliance on local knowledge, biodiversity conservation, and community involvement.

6.1 Traditional Farming Practices

6.1.1 Agrobiodiversity Conservation: The Resilience of Traditional Farming Systems

Traditional farming systems have long recognized the importance of agrobiodiversity, as they typically involve the cultivation of diverse crop varieties and the rearing of locally adapted animal breeds. This practice of embracing diverse agricultural resources has been shaped over generations to suit the specific conditions of the local environment (Altieri and Nicholls 2013). The emphasis on agrobiodiversity in traditional farming systems is in line with the principles of agroecology, which aims to enhance ecosystem resilience, improve soil health, and reduce dependency on external inputs.

Agrobiodiversity refers to the variety of plants, animals, and microorganisms that are utilized in agricultural systems. Traditional farmers have played a vital role in the conservation and utilization of agrobiodiversity, as they have developed and maintained a wide array of crop varieties and animal breeds that are adapted to their local conditions. This diversity serves as a valuable resource for food security and resilience, as different varieties and breeds possess varying traits that can help withstand environmental challenges such as pests, diseases, and climate fluctuations.

M. Otieno, *Sustainable Agroecological Practices in Sub-Saharan Africa in the Face of Climate Change*, Advances in Geographical and Environmental Sciences, https://doi.org/10.1007/978-3-031-70472-7_6

The incorporation of diverse crop varieties and animal breeds within traditional farming systems promotes biodiversity on multiple levels. On-farm diversity contributes to the overall diversity of species within agricultural landscapes, providing habitats and resources for beneficial organisms such as pollinators and natural enemies of pests. Furthermore, the presence of diverse crops can help break pest and disease cycles by reducing the buildup of pathogens and pests associated with particular plant species.

Agrobiodiversity also plays a crucial role in improving soil health within traditional farming systems. Different crop varieties have unique root structures and exudates, which can enhance soil microbial diversity and nutrient cycling. The diverse plant and animal components in traditional agroecosystems contribute to the maintenance of soil fertility, water-holding capacity, and overall soil health, thereby increasing the long-term sustainability of these systems.

Despite the recognized benefits of agrobiodiversity, traditional farming systems are under threat from the expansion of industrial or chemical agriculture, which often promotes monocultures and the use of external inputs. Furthermore, climate change poses an additional challenge to the conservation of agrobiodiversity, as it can disrupt the delicate balance of traditional farming systems.

There is an urgent need to adopt an agroecosystems approach that goes beyond the focus on genetic resource conservation alone. Strategies that build upon valuable local experiences and knowledge in traditional farming practices, while also taking advantage of recent scientific findings in agroecology and ecosystem health, can be instrumental in preserving the resilience of these systems.

6.1.2 Sustainable Soil Management in Traditional Farming Systems

Traditional farming practices have long recognized the importance of soil conservation and fertility enhancement as essential components of sustainable agricultural systems. These practices, which include crop rotation, intercropping, agroforestry, and the use of organic matter, have been widely adopted due to their ability to promote soil health, nutrient cycling, and water retention (Doran 2002; Kilbas et al. 2006).

Crop rotation is a common technique in traditional farming where different crops are grown in sequence on the same land (Doran 2002). This practice helps break pest and disease cycles, reduces soil erosion, and improves soil structure and nutrient availability. By alternating crops with varying nutrient requirements and growth characteristics, crop rotation supports balanced nutrient cycling and reduces the buildup of pests and pathogens in the soil.

Intercropping, where two or more crops are grown simultaneously on the same piece of land, promotes biodiversity within the agricultural system, enhances resource utilization, and improves soil health. Intercropping can lead to complementary interactions among crops, such as nutrient sharing and pest repellence, which contribute to increased overall productivity and improved soil conditions (Fontaine et al. 2024).

Agroforestry, the integration of trees with crops or livestock, is another prominent technique observed in traditional farming systems. The presence of trees provides numerous benefits, such as improved soil structure, increased organic matter input, enhanced water infiltration, and the provision of shade and wind protection. Agroforestry systems contribute to sustainable soil management by reducing soil erosion, supporting nutrient cycling, and enhancing soil moisture retention.

The use of organic matter, such as compost and manure, is also a critical component of traditional farming practices. Organic matter plays a vital role in maintaining and enhancing soil fertility by increasing the availability of essential nutrients, improving soil structure, and supporting a diverse and healthy soil microbial community (Kilbas et al. 2006; Swami 2020; Nayak et al. 2020).

6.1.3 Sustainable Water Management in Traditional Farming Systems

Traditional farming practices have long recognized the importance of water management techniques tailored to local conditions, aligning with the principles of agroecology. These techniques aim to promote efficient water use, minimize soil erosion, and support sustainable water management in agricultural systems (Khan and Hanjra 2008).

Terracing, a commonly employed traditional irrigation method in hilly or mountainous regions, involves constructing level or gently sloping platforms on slopes, creating steps or terraces that follow the contour of the land. Terraces help to reduce water runoff and soil erosion by slowing down the flow of water, allowing it to infiltrate into the soil. This technique effectively conserves water resources and enhances soil moisture retention, particularly in areas with steep slopes (Hudson 1987).

Contour farming is another traditional water management practice that is widely adopted in regions with undulating topography. It involves cultivating crops along the contour lines of the land, forming ridges and furrows that act as barriers to prevent runoff and promote water infiltration. Contour farming helps to retain soil moisture, minimize soil erosion caused by water runoff, and improve the efficiency of water use in agricultural fields (Hudson 1987; Rao 2005).

Water harvesting techniques, such as the construction of small-scale ponds or reservoirs, are commonly used in traditional farming systems to capture and store rainwater for later use in irrigation. These water harvesting structures enable farmers to collect and store water during the rainy season, providing a reliable source of irrigation water during dry periods (Gebreyess and Amare 2019). This practice helps to maintain crop productivity, particularly in regions where water availability is limited. By incorporating these traditional water management techniques, traditional farming systems demonstrate a holistic approach to sustainable water management. These practices not only enhance water-use efficiency but also promote soil and water conservation, contributing to the overall sustainability of agricultural systems (Hudson 1987; Rao 2005; Gebreyess and Amare 2019).

A community dam used to store water for irrigation

The effectiveness of these traditional water management practices has been widely documented. For example, studies have shown that water harvesting techniques can significantly improve crop yields in semi-arid regions with limited water availability (Stroosnijder et al. 2012; Gebreyess and Amare 2019). Furthermore, the adoption of soil and water conservation measures, such as terracing and contour farming, has been shown to increase soil moisture, reduce soil erosion, and enhance overall crop productivity (Hudson 1987; Stroosnijder et al. 2012; Gebreyess and Amare 2019; Kumawat et al. 2020).

Overall, the incorporation of traditional water management techniques in agricultural systems can play a crucial role in promoting sustainable water use, mitigating the impacts of water scarcity, and ensuring the long-term viability of farming practices.

6.1.4 The Importance of Community Participation and Knowledge Sharing in Sustainable Agriculture

Traditional farming systems have long recognized the significance of strong community participation, knowledge sharing, and collective decision-making as key social aspects of sustainable agriculture. These social dimensions are closely aligned with the principles of agroecology, which emphasizes the importance of local knowledge, farmer empowerment, and participatory approaches in developing sustainable agricultural systems.

In traditional farming systems, community participation plays a vital role in shaping agricultural practices and decision-making processes (Heryanto et al. 2018). Farmers actively engage with their local communities, sharing their knowledge, experiences, and innovative techniques (Herrera-Reyes et al. 2018; Heryanto et al. 2018; Ghosh et al. 2020). This collective participation fosters a sense of ownership and responsibility, as well as a deep understanding of local conditions, challenges, and opportunities (Coulibaly et al. 2021). Knowledge sharing is a cornerstone of traditional farming systems, where farmers pass down their expertise and accumulated wisdom from generation to generation. This knowledge is context-specific and tailored to local agroecological conditions, enabling farmers to adapt their practices to the local environment and foster sustainable production systems. Traditional knowledge encompasses various aspects, such as seed saving, crop management, pest and disease control, and soil fertility enhancement (Ghosh et al. 2020; Coulibaly et al. 2021).

Collective decision-making is another characteristic of traditional farming systems. Farmers often engage in collaborative processes, where decisions regarding agricultural practices, resource management, and community development are made collectively. This participatory approach ensures that diverse perspectives, experiences, and needs are considered, leading to more inclusive and contextually appropriate solutions. Agroecology recognizes the value of local knowledge, farmer empowerment, and participatory approaches, as these elements are essential for the development of sustainable agriculture that is responsive to the unique needs and conditions of local communities (Carlile and Garnett 2021; Sethuraman et al. 2021).

Research has shown that community participation and knowledge sharing in traditional farming systems contribute to the overall sustainability of the agricultural sector. By fostering a sense of ownership and responsibility, as well as a deep understanding of local conditions, these social dimensions enable farmers to adapt their practices to the local environment and develop resilient production systems (Gliessman 1995; Sabar 2012; Carlile and Garnett 2021). Additionally, the collective decision-making processes that characterize traditional farming systems ensure that diverse perspectives and needs are considered, leading to more inclusive and contextually appropriate solutions (Coulibaly et al. 2021).

6.1.5 Resilience of Traditional Farming Practices to Climate Variability

Traditional farming practices have demonstrated remarkable adaptability to local climatic conditions, including periods of drought, floods, and temperature fluctuations, by incorporating various strategies, e.g. crop diversification, minimal tillage, use of indigenous varieties, etc., that promote resilience in farming systems and align with the core principles of agroecology (Gupta 2010; Agbongiarhuoyi et al. 2013; Altieri and Nicholls 2013; Mubiru et al. 2018).

One key adaptation strategy employed in traditional farming systems is the cultivation of drought-resistant crop varieties. Farmers have developed and selected crop varieties that are adapted to their local environments, exhibiting traits such as drought tolerance, water-use efficiency, and resilience to climatic stresses (Gupta 2010). These locally adapted varieties often have diverse genetic backgrounds and possess inherent mechanisms that enable them to withstand periods of water scarcity and survive under challenging conditions (Gupta 2010). The preservation and selection of seeds from locally adapted crop varieties, known as the practice of seed saving, further contributes to the resilience of traditional farming systems by maintaining a pool of genetic resources capable of adapting to changing environmental conditions (Gupta 2010; Mubiru et al. 2018).

Water conservation techniques are also integral to traditional farming practices, particularly in regions prone to drought or water scarcity. Farmers employ various methods to optimize water use, such as mulching, contour bunding, and rainwater harvesting. Mulching involves the application of organic materials or crop residues to the soil surface, which reduces evaporation and helps retain soil moisture. Contour bunding, on the other hand, refers to the construction of soil bunds along the contour lines of the land, which prevents soil erosion and promotes water infiltration. Rainwater harvesting techniques, including the construction of ponds or terraces, capture and store rainfall for use during dry periods, reducing dependence on external water sources (Mubiru et al. 2018; Amenuvor 2019).

By leveraging local knowledge and traditional practices, farmers can enhance their capacity to cope with climate variability. The adaptation and mitigation potential of traditional agriculture in the face of a changing climate has been well-documented, as observed agricultural performance after extreme climatic events has revealed that resilience to climate disasters is closely linked to the high level of on-farm biodiversity, a typical feature of traditional farming systems (Gupta 2010; Altieri and Nicholls 2013; Mubiru et al. 2018; Amenuvor 2019). Traditional management systems combined with the use of agroecologically based management strategies may represent a viable and robust path to increase the productivity, sustainability, and resilience of peasant-based agricultural production under predicted climate scenarios.

6.1.6 The Synergistic Integration of Traditional Farming Practices and Agroecology

Traditional farming practices offer valuable insights and experiences that can significantly enhance the implementation of agroecological approaches. The integration of traditional knowledge with scientific advancements allows for the development of sustainable and resilient agricultural systems that prioritize environmental conservation, social equity, and food security.

Traditional farming systems have accumulated knowledge and expertise over generations, adapting to local conditions and addressing the challenges associated with agricultural production. This traditional knowledge encompasses a deep understanding of ecological interactions, local agroclimatic conditions, pest and disease dynamics, soil fertility management, and traditional agricultural practices (Thrupp 2000). These insights provide valuable guidance in the development and implementation of agroecological approaches (Altieri and Nicholls 2013; Sabar 2012; Carlile and Garnett 2021).

Agroecology recognizes the importance of integrating traditional knowledge into scientific research and agricultural practices. By combining traditional wisdom with scientific advancements, agroecology can benefit from the synergistic effects of traditional practices and innovative techniques. Traditional knowledge offers context-specific solutions that have proven effective in promoting sustainable and resilient agricultural systems. The integration of traditional knowledge into agroecology allows for the identification and utilization of locally adapted practices and techniques, such as strategies for soil conservation, water management, pest and disease control, and crop diversity.

These traditional practices have been shown to enhance agroecology's ability to address current and emerging challenges in sustainable agriculture (Thrupp 2000; Sabar 2012; Carlile and Garnett 2021; Fontaine et al. 2024). Furthermore, traditional farming practices prioritize environmental conservation, social equity, and food security (Thrupp 2000; Sabar 2012; Carlile and Garnett 2021; Fontaine et al. 2024). The blending of traditional knowledge and scientific approaches in agroecology can lead to the development of holistic and resilient agricultural systems that are better equipped to address the complex issues faced by contemporary agriculture (Gliessman 1995; Thrupp 2000; Sabar 2012; Uprety et al. 2012; Gliessman 2020; Carlile and Garnett 2021; Sethuraman et al. 2021; Fontaine et al. 2024).

6.2 Industrial Farming Practices

6.2.1 Contrasting Industrial Farming and Agroecological Practices: Toward Sustainable and Regenerative Agriculture

Industrial farming practices, often associated with conventional agriculture, differ significantly from the principles and practices of agroecology. While industrial farming prioritizes high-input, large-scale production systems, agroecology promotes sustainable and regenerative practices that prioritize ecological integrity, social equity, and long-term viability (Rajeshwari and Nagarajan 2016; Lucas 2021; Veldkamp et al. 2023).

One key difference lies in the reliance on external inputs (Rajeshwari and Nagarajan 2016; Sethuraman et al. 2021). Industrial farming practices heavily depend on synthetic fertilizers, pesticides, and genetically modified organisms (GMOs) to achieve maximum crop yields (Rajeshwari and Nagarajan 2016). These practices can lead to nutrient imbalances, soil degradation, and water pollution due to excessive nutrient runoff. In contrast, agroecology emphasizes the use of organic fertilizers, such as compost, manure, and green manure cover crops, which enhance soil fertility, improve nutrient availability, and promote long-term soil health (Veldkamp et al. 2023).

Furthermore, industrial agriculture's heavy use of synthetic pesticides can have detrimental effects on ecosystems, including the loss of biodiversity, negative impacts on beneficial organisms, and the development of pesticide resistance. Agroecology, on the other hand, seeks to minimize the reliance on external inputs by emphasizing natural and organic methods that capitalize on ecological processes such as nutrient cycling, pest management, and soil health improvement (Altieri 1999; Sethuraman et al. 2021).

Studies have shown that agroforestry, a form of agroecological management, can outperform open cropland and grassland in terms of multifunctionality, providing important ecosystem functions and landscape features that stimulate biodiversity and carbon sequestration while reducing environmental pollution and soil degradation (Veldkamp et al. 2023). Transformative agroecology has been recognized as a stepping stone to achieving several Sustainable Development Goals, due to its great potential to build climate change-resilient farming systems while enhancing ecosystem services and reducing biodiversity loss (Gliessman 1995; Rajeshwari and Nagarajan 2016; Carlile and Garnett 2021; Sethuraman et al. 2021; Veldkamp et al. 2023).

6.2.2 Monoculture Versus Polyculture: Evaluating the Ecological and Sustainability Implications

Industrial farming has long relied on monoculture systems, which involve the cultivation of a single crop over vast expanses of land. In contrast, agroecology emphasizes the use of polyculture and diversification, integrating multiple crops and sometimes livestock within the same farming system. Polyculture offers numerous benefits, including enhanced biodiversity, reduced risks of pests and diseases, improved soil health, and the promotion of ecological balance.

A tea plantation in eastern Kenya

Monoculture systems, prevalent in industrial farming, prioritize the cultivation of a single crop species over large areas. This practice facilitates the standardization of production processes, simplifies management practices, and maximizes efficiency in terms of input application and harvesting (Sethuraman et al. 2021). However, monocultures can have adverse environmental consequences. The absence of crop diversity reduces habitat availability for beneficial organisms, disrupts ecological balances, and increases vulnerability to pest and disease outbreaks (Rajeshwari and Nagarajan 2016). Monocultures also often require intensive use of synthetic inputs, such as pesticides and fertilizers, which can lead to environmental degradation and negative impacts on human health (Rajeshwari and Nagarajan 2016).

Agroecology, on the other hand, emphasizes the importance of polyculture and diversification in farming systems. Polyculture involves the cultivation of multiple crop species within the same field or farm, often incorporating intercropping, crop rotation, and companion planting techniques (Rajeshwari and Nagarajan 2016). These practices promote biodiversity, both above and below ground, by creating diverse habitats and supporting a wider range of plant and animal species. Diverse plant communities provide niches for beneficial insects, pollinators, and natural enemies of pests, contributing to natural pest control and reducing the reliance on synthetic pesticides (Rajeshwari and Nagarajan 2016; Forister et al. 2019; Sethuraman et al. 2021; Veldkamp et al. 2023).

The benefits of polyculture-based agroecology extend beyond ecological considerations. Diverse farming systems can also enhance the resilience of agricultural production in the face of climate change and other environmental stresses. Additionally, agroforestry, which integrates trees and crops within the same system, has been shown to outperform open cropland and grassland in providing a range of ecosystem services (Veldkamp et al. 2023).

While industrial monoculture farming prioritizes productivity and profitability, the long-term environmental and social costs associated with this approach are increasingly recognized. Agroecology, with its emphasis on polyculture and diversification, offers a more sustainable alternative that can meet the growing demand for food while preserving ecosystem health and resilience.

6.2.3 Agroecology: Promoting Soil Health and Sustainability

Soil health and conservation are critical components of sustainable agriculture, as industrial farming practices often contribute to soil degradation, erosion, and nutrient depletion. In contrast, agroecology emphasizes the importance of maintaining soil fertility, structure, and biodiversity through various practices such as cover cropping, crop rotation, agroforestry, and the addition of organic matter (Hurni et al. 2015; Carlile and Garnett 2021; Sethuraman et al. 2021; Veldkamp et al. 2023).

Intensive tilling, a common practice in industrial farming, can disrupt the natural structure and composition of the soil, leading to increased soil erosion and loss of organic matter. The repeated disturbance of the soil through tillage can break down soil aggregates, reducing soil porosity and increasing vulnerability to erosion by wind and water. Furthermore, the use of heavy machinery in industrial farming can lead to soil compaction, further compromising soil structure and limiting root growth. Agroecology advocates for reduced tillage or no-till practices, which minimize soil disturbance, preserve soil structure, and help mitigate erosion (Tamburini et al. 2016).

Chemical inputs, such as synthetic fertilizers and pesticides, commonly used in industrial farming, can also have detrimental effects on soil health. Excessive fertilizer application can disrupt nutrient cycling, leading to nutrient imbalances and depletion of essential micronutrients over time. Prolonged reliance on synthetic fertilizers can also reduce soil microbial activity and diversity, impacting nutrient availability and soil ecosystem functions. Agroecological practices, such as cover cropping and crop rotation, enhance nutrient cycling and reduce the need for external inputs by incorporating nitrogen-fixing plants and diverse crop species that have different nutrient requirements (Altieri 1999; Tamburini et al. 2016; Barman et al. 2019; Veldkamp et al. 2023).

Multifunctional agroforestry systems, which integrate trees and shrubs with agricultural crops, have been shown to outperform open cropland and grassland in providing various ecosystem services, including improved soil health and conservation (Veldkamp et al. 2023). The integration of trees and diverse plant species in agroforestry systems can enhance soil organic matter, nutrient cycling, and microbial activity, contributing to long-term soil fertility and resilience.

The use of beneficial microorganisms, such as plant-growth-promoting rhizobacteria and mycorrhizal fungi, has also been explored as a promising approach to enhance soil health and reduce the reliance on synthetic inputs. These microorganisms can improve nutrient availability, suppress soil-borne pathogens, and enhance plant growth, supporting more sustainable and productive agricultural systems (Altieri 1999; Barman et al. 2019; Menalled et al. 2021; Fontaine et al. 2024; Veldkamp et al. 2023).

Agroecological practices that prioritize soil health and conservation, such as reduced tillage, cover cropping, crop rotation, agroforestry, and the incorporation of beneficial microorganisms, offer a promising path toward more sustainable and resilient agricultural systems (Barman et al. 2019; Bolinder et al. 2020; Tahat et al. 2020; González-Pérez et al. 2022; Fontaine et al. 2024; Veldkamp et al. 2023). These nature-based solutions not only contribute to improved soil fertility and ecosystem functions but also help mitigate the environmental impacts of industrial farming.

6.2.4 Sustainable Water Management in Agroecological Farming

The intensification of industrial farming practices has led to significant challenges in water management, often resulting in water wastage, depletion of water resources, and water pollution. Conversely, agroecology offers alternative techniques that prioritize water conservation and sustainable water management, contributing to the overall sustainability of agricultural systems.

Industrial farming commonly relies on extensive irrigation systems, which can be highly inefficient and lead to significant water losses. Excessive irrigation can result in runoff and percolation, causing water to be lost beyond the root zone of crops and depleting local water resources (Awais et al. 2020; Nhemachena et al. 2020). Additionally, the application of synthetic fertilizers and pesticides in these systems can contribute to water pollution, as these chemicals can leach into groundwater or run off into nearby water bodies, posing ecological and human health risks.

In contrast, agroecological approaches recognize the importance of sustainable water management and offer alternative practices to address these challenges. Rainwater harvesting techniques, such as the use of catchment systems or small reservoirs, allow farmers to collect and store rainwater during periods of high rainfall for later use during dry periods, reducing reliance on external water sources and minimizing water wastage (Tewari and Tewari 2003). Efficient irrigation methods, such as drip irrigation and precision sprinkler systems, are also promoted in agroecology, as they deliver water directly to the root zone of plants, reducing evaporation losses and ensuring more effective water use (Tewari and Tewari 2003; García-Garizábal et al. 2012; Awais et al. 2020).

By optimizing water distribution and promoting water-saving practices, these techniques contribute to the conservation of water resources and increased irrigation efficiency (Tewari and Tewari 2003; García-Garizábal et al. 2012). Efficient water resource management is crucial for achieving food security and sustaining livelihoods, and agroecology provides a holistic approach to addressing these complex water-related challenges.

6.2.5 The Role of Agroecology in Preserving Biodiversity and Habitat Conservation

Industrial farming practices, characterized by extensive land clearing and intensive agrochemical use, often have detrimental impacts on biodiversity, leading to habitat loss and ecosystem fragmentation (Thrupp 2000). In contrast, agroecology employs various strategies to promote biodiversity and support beneficial organisms within agricultural landscapes (Batáry et al. 2011; Veldkamp et al. 2023).

Land clearing for industrial monoculture farming can result in the destruction of natural habitats and a significant loss of biodiversity (Thrupp 2000; Batáry et al. 2011). Large-scale monoculture systems offer limited resources and habitats for diverse plant and animal species. Moreover, the widespread use of synthetic fertilizers and pesticides can have detrimental effects on non-target organisms, such as pollinators, beneficial insects, and soil microorganisms, leading to their population decline and reduced biodiversity (Thrupp 2000; Batáry et al. 2011).

Agroecology offers a different approach by implementing practices that preserve and enhance biodiversity. Habitat restoration and the creation of wildlife-friendly areas within agricultural landscapes are key strategies. By establishing hedgerows, field margins, and buffer zones, farmers can provide habitats for a diverse range of organisms, including birds, mammals, insects, and beneficial microorganisms, which act as refuges and support ecosystem services (Batáry et al. 2011; Veldkamp et al. 2023). Agroforestry systems, which integrate trees or shrubs with crops, also play a vital role in promoting biodiversity within agricultural landscapes (Veldkamp et al. 2023).

Agroecological farming practices, such as reducing the use of agrochemicals and adopting integrated pest management, can help to mitigate the negative impacts of industrial agriculture on biodiversity. By fostering a mosaic of diverse crops and habitats, agroecology can support a wide range of species and maintain the ecological balance necessary for sustainable agricultural production and ecosystem resilience.

6.2.6 The Social and Economic Benefits of Agroecology

Industrial farming systems have long prioritized centralized control and decision-making, often at the expense of local communities. This has led to social inequalities, rural displacement, and economic concentration (Lucas 2021; Sethuraman et al. 2021). In contrast, agroecology places a strong emphasis on community participation, farmer empowerment, knowledge sharing, and the equitable distribution of resources.

The displacement of small-scale farmers and the consolidation of land into larger, corporate-owned operations have been a common outcome of industrial farming practices. This concentration of power and resources can contribute to social inequalities and economic disparities, as local communities lose control over their agricultural practices and livelihoods. Agroecology, on the other hand, aims to empower farmers and local communities by involving them in decision-making processes, fostering collective action, and supporting local food systems.

Agroecology recognizes the value of local knowledge and traditional farming practices and emphasizes the importance of knowledge sharing and farmer-to-farmer exchange. By promoting participatory approaches, agroecology enables farmers to be active participants in the design and implementation of sustainable agricultural systems. This involvement not only leads to more context-appropriate solutions but also empowers farmers to adapt and innovate in response to local challenges and changing conditions (Carlile and Garnett 2021; Gliessman 2020).

Furthermore, agroecology supports the equitable distribution of resources, including land, water, and seeds. By promoting agrobiodiversity and seed sovereignty, agroecology challenges the dominance of industrial agricultural systems, which often rely on patented and genetically modified inputs that can limit the autonomy and self-determination of local communities.

References

Agbongiarhuoyi AE, Abdulkarim IF, Fawole OP, Obatolu BO, Famuyiwa BS, Oloyede A (2013, July 4) Analysis of farmers' adaptation strategies to climate change in cocoa production in Kwara State. J Agric Exten 17(1):10–10. https://doi.org/10.4314/jae.v17i1.2

Altieri MA (1999) The ecological role of biodiversity in agroecosystems. Agric Ecosyst Environ 74(1–3):19–31. https://doi.org/10.1016/s0167-8809(99)00028-6

Altieri MA, Nicholls CI (2013) The adaptation and mitigation potential of traditional agriculture in a changing climate. Clim Chang 140(1):33–45. https://doi.org/10.1007/s10584-013-0909-y

Amenuvor FK (2019, October 31) Climate change adaptation strategies adopted by smallholder farmers and its effects on food security at Anloga in the Volta Region of Ghana, West Africa. Int J Human Social Stud 7(10). https://doi.org/10.24940/theijhss/2019/v7/i10/hs1910-093

Awais M, Li W, Yang YF, Ji LL (2020) Detection of plant water stress using UAV thermal images for precision farming application. Appl Ecol Environ Res 18(3):4087–4102. https://doi.org/10.15666/aeer/1803_40874102

Barman S, Das S, Bhattacharya SS (2019) The prospects of bio-fertilizer technology for productive and sustainable agricultural growth. In: New and future developments in microbial biotechnology and bioengineering. Elsevier, pp 233–253. https://doi.org/10.1016/b978-0-444-64191-5.00017-1

Batáry P, Báldi A, Kleijn D, Tscharntke T (2011) Landscape-moderated biodiversity effects of agri-environmental management: a meta-analysis. Proc R Soc B Biol Sci 278(1713):1894–1902. https://doi.org/10.1098/rspb.2010.1923

Bolinder MA, Crotty F, Elsen A, Frac M, Kismányoky T, Lipiec J et al (2020) The effect of crop residues, cover crops, manures and nitrogen fertilization on soil organic carbon changes in agroecosystems: a synthesis of reviews. Mitig Adapt Strateg Glob Chang 25:929–952. https://doi.org/10.1007/s11027-020-09916-3

Carlile R, Garnett T (2021) What is agroecology. TABLE Explainer Series. TABLE, University of Oxford, Swedish University of Agricultural Sciences and Wageningen University & Research

Coulibaly PT, Du J, Dabuo FT, Akouatcha GH (2021, January 1) The effect of farmers social networks on sustainable agricultural practices adoption: a scoping review protocol. Int J 8(1):56–60. https://doi.org/10.18488/journal.70.2021.81.56.60
Doran JW (2002) Soil health and global sustainability: translating science into practice. Agric Ecosyst Environ 88(2):119–127. https://doi.org/10.1016/s0167-8809(01)00246-8
Fontaine S, Abbadie L, Aubert M, Barot S, Bloor JM, Derrien D et al (2024) Plant–soil synchrony in nutrient cycles: learning from ecosystems to design sustainable agrosystems. Glob Chang Biol 30(1):e17034. https://doi.org/10.1111/gcb.17034
Forister ML, Pelton EM, Black SH (2019) Declines in insect abundance and diversity: we know enough to act now. Conserv Sci Pract 1(8):e80. https://doi.org/10.1111/csp2.80
García-Garizábal I, Abrahao R, Causapé J (2012) Watershed monitoring for the assessment of irrigation water use and irrigation contamination. In: Irrigation and water management, pollution and alternative strategies, pp 21–38. https://doi.org/10.5772/31455
Gebreyess BF, Amare A (2019) Water harvesting technologies in semi-arid and arid areas. J Degraded Mining Lands Manag 7(1):1921. https://doi.org/10.15243/jdmlm.2019.071.1921
Ghosh MK, Hasan S, EnamulHaque M, JamalUddin M (2020, January 25) Knowledge of farmers to sustainable agriculture practices: a case study in southwestern region of Bangladesh. Scholars J Agric Vet Sci 07(01):5–12. https://doi.org/10.36347/sjavs.2020.v07i01.002
Gliessman SR (1995) 3 Sustainable agriculture: an agroecological perspective. In: Advances in plant pathology, vol 11. Academic Press, pp 45–57. https://doi.org/10.1016/s0736-4539(06)80005-x
Gliessman SR (2020) Transforming food and agriculture systems with agroecology. Agric Hum Values 37:547–548. https://doi.org/10.1007/s10460-020-10058-0
González-Pérez E, Ortega-Amaro MA, Bautista E, Delgado-Sánchez P, Jiménez-Bremont JF (2022) The entomopathogenic fungus Metarhizium anisopliae enhances Arabidopsis, tomato, and maize plant growth. Plant Physiol Biochem 176:34–43. https://doi.org/10.1016/j.plaphy.2022.02.008
Gupta J (2010) A history of international climate change policy. Wiley Interdiscip Rev Clim Chang 1(5):636–653. https://doi.org/10.1002/wcc.67
Herrera-Reyes AT, Carmenado IDLR, Martínez-Almela J (2018) Project-based governance framework for an agri-food cooperative. Sustain For 10(6):1881. https://doi.org/10.3390/su10061881
Heryanto MA, Supyandi D, Sukayat Y (2018, June) Agricultural local knowledge system: lesson learn from rural development in Mekarjaya Village, Kabupaten Purwakarta. In: IOP conference series: earth and environmental science, vol 166, No. 1. IOP Publishing, p 012046. https://doi.org/10.1088/1755-1315/166/1/012046
Hudson N (1987) Soil and water conservation in semi-arid areas (No 57). Food Agric Org https://www.amazon.com/Water-Conservation-Semi-arid-Areas-bulletins/dp/9251026068
Hurni H, Giger M, Liniger H, Studer RM, Messerli P, Portner B et al (2015) Soils, agriculture and food security: the interplay between ecosystem functioning and human well-being. Curr Opin Environ Sustain 15:25–34. https://doi.org/10.1016/j.cosust.2015.07.009
Khan S, Hanjra MA (2008) Sustainable land and water management policies and practices: a pathway to environmental sustainability in large irrigation systems. Land Degrad Dev 19(5):469–487. https://doi.org/10.1002/ldr.852
Kilbas AA, Srivastava HM, Trujillo JJ (2006) Theory and applications of fractional differential equations, vol 204. Elsevier. https://doi.org/10.1016/s0304-0208(06)80001-0
Kumawat A, Yadav D, Samadharmam K, Rashmi I (2020) Soil and water conservation measures for agricultural sustainability. Soil Moist Import 23. https://doi.org/10.5772/intechopen.92895
Lucas V (2021) A "silent" agroecology: the significance of unrecognized sociotechnical changes made by French farmers. Rev Agric Food Environ Stud 102(1):1–23. https://doi.org/10.1007/s41130-021-00140-4
Menalled UD, Seipel T, Menalled FD (2021) Farming system effects on biologically mediated plant–soil feedbacks. Renew Agric Food Syst 36(1):1–7. https://doi.org/10.1017/s1742170519000528

Mubiru DN, Radeny M, Kyazze FB, Zziwa A, Lwasa J, Kinyangi J, Mungai C (2018) Climate trends, risks and coping strategies in smallholder farming systems in Uganda. Clim Risk Manag 22:4–21. https://doi.org/10.1016/j.crm.2018.08.004

Nayak B, Rath BS, Shahid M, Jena SN, Bagchi TB, Roy PS (2020) Organic nutrient management in aromatic rice-rice sequence: a critical review. Int J Chem Stud 8:1435–1444. https://doi.org/10.22271/chemi.2020.v8.i5t.10503

Nhemachena C, Nhamo L, Matchaya G, Nhemachena CR, Muchara B, Karuaihe ST, Mpandeli S (2020) Climate change impacts on water and agriculture sectors in Southern Africa: threats and opportunities for sustainable development. Water 12(10):2673. https://doi.org/10.3390/w12102673

Rajeshwari S, Nagarajan M (2016) The role of modern technology in agribusiness management. J Manag Sci 6(3):320–325. https://doi.org/10.26524/jms.2016.33

Rao RJ (2005) Participatory watershed management (PWM): an approach for integrated development of rural India: a case study from Karnataka, southern India. Int J Environ Technol Manag 5(1):107–115. https://doi.org/10.1504/ijetm.2005.006510

Sabar B (2012) Bringing culture back: traditional agricultural knowledge, food production and sustainable livelihood among Chuktia Bhunjia of Orissa. Social Change 42(2):203–227. https://doi.org/10.1177/004908571204200204

Sethuraman G, Mohd Zain NA, Yusoff S, Ng YM, Baisakh N, Cheng A (2021) Revamping ecosystem services through agroecology—the case of cereals. Agriculture 11(3):204. https://doi.org/10.3390/agriculture11030204

Stroosnijder L, Moore D, Alharbi A, Argaman E, Biazin B, van den Elsen E (2012) Improving water use efficiency in drylands. Curr Opin Environ Sustain 4(5):497–506. https://doi.org/10.1016/j.cosust.2012.08.011

Swami S (2020) Soil health management under organic production system. IJCS 8(2):330–339. https://doi.org/10.22271/chemi.2020.v8.i2e.8789

Tahat MM, Alananbeh KM, Othman YA, Leskovar DI (2020) Soil health and sustainable agriculture. Sustain For 12(12):4859. https://doi.org/10.3390/su12124859

Tamburini G, De Simone S, Sigura M, Boscutti F, Marini L (2016) Soil management shapes ecosystem service provision and trade-offs in agricultural landscapes. Proc R Soc B Biol Sci 283(1837):20161369. https://doi.org/10.1098/rspb.2016.1369

Tewari A, Tewari S (2003) Water resource management in India: issues and options. Indian J Public Admin 49(3):411–420. https://doi.org/10.1177/0019556120030318

Thrupp LA (2000) Linking agricultural biodiversity and food security: the valuable role of agrobiodiversity for sustainable agriculture. Int Aff 76(2):265–281. https://doi.org/10.1111/1468-2346.00133

Uprety Y, Asselin H, Bergeron Y, Doyon F, Boucher JF (2012) Contribution of traditional knowledge to ecological restoration: practices and applications. Ecoscience 19(3):225–237. https://doi.org/10.2980/19-3-3530

Veldkamp E, Schmidt M, Markwitz C, Beule L, Beuschel R, Biertümpfel A et al (2023) Multifunctionality of temperate alley-cropping agroforestry outperforms open cropland and grassland. Commun Earth Environ 4(1):20. https://doi.org/10.1038/s43247-023-00680-1

Chapter 7
Biodiversity and Ecosystem Services

Biodiversity and ecosystem services are fundamental components of agroecology, and the principles and practices of agroecology aim to promote and enhance them. Agroecology recognizes the critical role of biodiversity in agricultural systems and emphasizes the importance of ecosystem services in supporting sustainable food production.

7.1 Biodiversity Conservation: The Cornerstone of Agroecological Resilience

Agroecology recognizes the intrinsic value of biodiversity in agricultural landscapes and emphasizes the importance of conserving and promoting diverse plant and animal species (Thrupp 2000; Rashid et al. 2021). This includes safeguarding crop genetic diversity, preserving wild relatives of cultivated plants, and supporting the presence of beneficial organisms such as pollinators, natural predators, and soil microorganisms. By prioritizing biodiversity conservation, agroecological systems contribute to ecological balance, enhance resilience, and promote long-term sustainability (Altieri 1999; Batáry et al. 2011).

Conservation of crop genetic diversity is a fundamental aspect of agroecology (Thrupp 2000). Traditional farming systems have long embraced diverse crop varieties that have been adapted to local conditions over generations (Altieri 1999). Agroecology builds on this approach and recognizes the value of maintaining a wide range of crop genetic resources, which can provide resilience in the face of environmental challenges such as pests, diseases, and climate change (Altieri 1999; Thrupp 2000). Conserving crop genetic diversity also ensures the availability of diverse traits that can be utilized for breeding programs, contributing to the

M. Otieno, *Sustainable Agroecological Practices in Sub-Saharan Africa in the Face of Climate Change*, Advances in Geographical and Environmental Sciences, https://doi.org/10.1007/978-3-031-70472-7_7

development of more resilient and adaptable crop varieties (Nodari and Guerra 2015; Thrupp 2000).

In addition to crop diversity, agroecology promotes the preservation of wild relatives of cultivated plants. Wild relatives often possess valuable genetic traits that can be used to enhance crop resilience, productivity, and nutritional quality. By conserving wild relatives, agroecological systems maintain a genetic pool that can be tapped into for crop improvement and adaptation to changing environmental conditions (Gill et al. 2014; Altieri 1999).

Agroecology also recognizes the vital role of beneficial organisms in maintaining ecological balance and promoting sustainable agricultural systems. Pollinators, such as bees, butterflies, and birds, are essential for the reproduction of many flowering plants (Altieri 1999; Thrupp 2000; Nodari and Guerra 2015). They contribute to increased crop yields and the production of nutritious foods. Natural predators, such as birds, bats, and insects, help to regulate populations of agricultural pests, reducing the need for harmful pesticides. Soil microorganisms, including fungi, bacteria, and nematodes, play a crucial role in nutrient cycling, soil structure, and plant health. By supporting the diversity of these beneficial organisms, agroecological systems enhance ecosystem services and create more resilient and self-regulating agricultural landscapes (Altieri 1999; Batáry et al. 2011; Nodari and Guerra 2015; Carlile and Garnett 2021).

The holistic approach of agroecology recognizes the interconnectedness of biodiversity, ecosystem services, and agricultural productivity. By prioritizing the conservation of biodiversity, agroecological systems contribute to the long-term sustainability of food production and the resilience of rural communities in the face of environmental challenges (Gliessman 1995, 2020; Altieri 1999; Gill et al. 2014; Erisman et al. 2016; Carlile and Garnett 2021; Sethuraman et al. 2021).

7.1.1 The Role of Agroecological Practices in Supporting Pollination and Pest Control

Maintaining biodiversity, particularly pollinator species like bees and butterflies, is crucial for effective pollination and agricultural productivity. Agroecological approaches prioritize the conservation and promotion of biodiversity to support pollinator populations and enhance pollination services, contributing to the sustainability of agricultural systems (Deutsch et al. 2018; Stout and Dicks 2022).

A key agroecological practice is the cultivation of diverse flowering plants, which provides a continuous and varied source of nectar and pollen for pollinators (Deutsch et al. 2018). This diversity of food resources sustains pollinator populations and ensures their presence for efficient crop pollination (Deutsch et al. 2018). Creating habitat corridors within and around agricultural landscapes is another important strategy to support pollinators by connecting fragmented habitats and providing refuge, nesting sites, and forage resources (Kearns et al. 1998).

Reducing the use of chemical pesticides is also a crucial aspect of agroecology in supporting pollinators. Many pesticides can have detrimental effects on pollinator populations, leading to declines in their numbers and diversity. Agroecological practices prioritize alternative pest management strategies like biological control and integrated pest management, which minimize reliance on harmful chemicals (Deutsch et al. 2018; Forister et al. 2019).

A bee visiting sunflower

By adopting these agroecological practices, agricultural systems can create safer environments for pollinators and promote their health and abundance, thereby enhancing pollination services and contributing to the productivity and sustainability of the system (Kearns et al. 1998; Potts et al. 2010; Deutsch et al. 2018; Forister et al. 2019). While honeybees remain critical for crop pollination, the conservation of native pollinator species is also essential (Potts et al. 2010) as their populations face threats from habitat fragmentation, changes in land use, and the spread of non-native species (Kearns et al. 1998). Furthermore, insect declines in general, including pollinators, can have far-reaching consequences for ecosystems and agricultural production (Potts et al. 2010; Deutsch et al. 2018; Forister et al. 2019). As the climate changes, the need for effective pollination and integrated pest management strategies will only increase, underscoring the importance of agroecological approaches (Deutsch et al. 2018).

7.1.2 Soil Health and Nutrient Cycling: The Importance of Agroecological Practices

Biodiversity in the soil, encompassing various organisms such as earthworms, fungi, bacteria, and other microorganisms, plays a crucial role in maintaining soil health and nutrient cycling. Agroecological practices prioritize the promotion and enhancement of soil biodiversity to support sustainable agricultural systems and improve soil fertility.

Cover cropping is an agroecological practice that involves planting specific crops during fallow periods or alongside cash crops to cover the soil surface. Cover crops help improve soil biodiversity by providing habitat and food sources for soil organisms and contribute to nutrient cycling by taking up excess nutrients from the soil and returning them when the cover crop is terminated and incorporated into the soil. The diverse root systems of cover crops further enhance soil structure, water infiltration, and carbon sequestration, leading to improved overall soil health (Andrew et al. 2012).

Crop rotation is another important agroecological practice that promotes soil biodiversity and nutrient cycling. Rotating different crops within a field over time helps break pest and disease cycles, reduces the buildup of specific pests and pathogens, and improves nutrient availability. Each crop in the rotation has unique root exudates and residues that support specific microbial communities, thereby increasing the diversity of soil organisms and their functions. This diversity enhances nutrient cycling processes and the availability of essential nutrients for plant uptake.

The addition of organic matter is a fundamental component of agroecological practices and is beneficial for soil biodiversity and nutrient cycling. Organic matter, such as compost, manure, or crop residues, serves as a food source for soil organisms, supporting their growth and activity (Nayak et al. 2020).

7.1.3 Water Management and Regulation: The Role of Biodiversity in Agroecological Systems

Biodiversity in and around water bodies, such as wetlands and riparian zones, plays a crucial role in water regulation, filtration, and purification. Agroecological approaches that prioritize the protection and restoration of these ecosystems can help maintain water quality, prevent erosion, and regulate water flow, benefiting both agricultural systems and downstream ecosystems.

Wetlands are recognized for their capacity to regulate water quantity and quality. They act as natural sponges, storing excess water during periods of heavy rainfall and gradually releasing it during drier periods, thus reducing the risk of flooding and drought. Wetlands also serve as effective filters, trapping sediments, nutrients, and pollutants from agricultural runoff, thereby improving water quality. The diverse vegetation and microorganisms found in wetlands contribute to the breakdown and removal of contaminants, further enhancing water purification processes (Rai et al. 2011).

Similarly, riparian zones, which are the areas adjacent to water bodies, play a vital role in maintaining water quality and regulating water flow. The vegetation in riparian zones helps stabilize riverbanks, reducing soil erosion and preventing sediments and pollutants from entering water bodies. The root systems of riparian plants enhance water infiltration, reducing the risk of surface runoff and promoting groundwater recharge. Furthermore, the diverse plant species in riparian zones provide shade, which helps regulate water temperature and create favorable conditions for aquatic organisms.

Agroecological approaches prioritize the protection and restoration of wetlands and riparian zones within agricultural landscapes to enhance water regulation, filtration, and purification. By implementing buffer strips of native vegetation along water bodies, farmers can reduce the amount of contaminants reaching these sensitive ecosystems, while also promoting a better balance between agricultural water use and the needs of wildlife (Lemly et al. 1993). Sustainable agriculture in wetland areas is an effective way to achieve equilibrium between wetland protection and agricultural use, ensuring the long-term viability of these vital ecosystems (Lemly et al. 1993; Rai et al. 2011; Zhu et al. 2013).

7.1.4 The Role of Biodiverse Agroecosystems in Climate Regulation and Resilience

Biodiverse agroecosystems play a crucial role in mitigating climate change and enhancing resilience to its impacts. The presence of diverse plant species within these systems contributes to carbon capture and storage, assisting in reducing greenhouse gas concentrations in the atmosphere (Rashid et al. 2021). Agroforestry practices, which integrate trees and crops, offer additional benefits such as providing shade, acting as windbreaks, and further enhancing carbon sequestration (Altieri and Nicholls 2013).

The process of carbon sequestration in biodiverse agroecosystems occurs through photosynthesis, where plants convert carbon dioxide into organic matter and store it in biomass and soils. Research has shown that these diverse agricultural systems have a higher potential for carbon storage compared to monoculture systems. Agroforestry, in particular, leverages the ability of trees to accumulate carbon above and below ground, thus increasing the overall carbon storage capacity of the agroecosystem.

Biodiversity within agroecosystems also enhances resilience to climate variability. Diverse plant species provide a wider range of functional traits and genetic variability, enabling them to better adapt to changing climatic conditions (Altieri and Nicholls 2013; Koskey et al. 2021; Rashid et al. 2021). For example, a diverse crop rotation system can include varieties with different tolerances to drought, floods, or temperature extremes, ensuring more stable yields across variable weather patterns.

Furthermore, diverse agroecosystems can better withstand and recover from extreme weather events, such as storms, floods, and droughts, due to the complementarity of different functional traits and the presence of natural buffers like trees and hedges (Batish et al. 2007; Rashid et al. 2021). Agroforestry, in particular, can play a crucial role in enhancing the resilience of agricultural systems, as the integration of trees provides shade, acts as windbreaks, and contributes to overall ecosystem stability (Koskey et al. 2021).

Given the increasing challenges from climate change, utilizing the adaptation and mitigation potential of biodiverse agroecosystems and agroforestry practices offers a promising strategy for creating more sustainable and resilient agricultural systems.

7.2 On-farm Biodiversity

On-farm biodiversity is a key component of agroecology, which recognizes the importance of diverse plant and animal species within agricultural systems. On-farm biodiversity refers to the presence and management of various species, including crops, livestock, insects, birds, and other organisms, within agricultural landscapes.

7.2.1 Harnessing Crop Diversity: The Agroecological Approach

Agroecology, an emerging paradigm in sustainable agriculture, emphasizes the cultivation of diverse crop species and varieties as a fundamental principle, encompassing the use of traditional, locally adapted cultivars, as well as heirloom an indigenous crop (Carlile and Garnett 2021). This approach offers numerous benefits, including risk reduction of crop failure, increased resilience to pests and diseases, and enhanced nutritional diversity (Gill et al. 2014; Francis et al. 2003; Carlile and Garnett 2021).

Diverse cropping systems can reduce the risk of crop failure caused by pests, diseases, and adverse weather conditions (Smith et al. 2023). By planting a mixture of crops, farmers can mitigate the likelihood of a single pest or disease devastating the entire harvest, as different crops have varying susceptibility to specific pests or diseases. Additionally, diverse cropping systems disrupt pest and disease cycles, reducing the buildup of populations and minimizing the need for synthetic pesticides (Sethuraman et al. 2021; Smith et al. 2023).

Crop diversity in agroecological systems also contributes to enhanced nutritional diversity, ensuring a more balanced and nutritious diet. Traditional and indigenous crops often possess unique nutritional profiles and offer a wider range of vitamins, minerals, and phytochemicals compared to modern monoculture crops (Gill et al. 2014; Jat et al. 2021). By promoting the cultivation of diverse crops, agroecology supports food systems that provide a broader array of nutrients, addressing the issue of malnutrition and enhancing food security (Gill et al. 2014; Jat et al. 2021; Sethuraman et al. 2021; Smith et al. 2023).

Several practices are employed in agroecological systems to promote crop diversity, such as interplanting, crop rotation, and mixed cropping. Interplanting, where different crops are grown together in the same field, allows for resource complementarity and increased productivity. Crop rotation, the successive cultivation of different crops on the same land, can enhance soil health, disrupt pest and disease cycles, and increase yields (Sethuraman et al. 2021; Smith et al. 2023).

The cultivation of diverse crop species and varieties is essential in agroecology, as it has the potential to enhance the resilience of agricultural systems. Research has shown that diverse cropping systems can reduce the risk of crop failure and enhance nutritional diversity.

7.2.2 Preserving Livestock Diversity for Sustainable Agriculture

Agroecology recognizes the paramount significance of diverse livestock breeds that are well-adapted to local environments and farming systems. Indigenous and traditional livestock breeds often possess valuable traits such as disease resistance, tolerance to harsh conditions, and efficient resource utilization, which are crucial for maintaining resilient and sustainable agricultural practices (Edwards et al. 1993; Gill et al. 2014).

The preservation and promotion of diverse livestock breeds are prioritized in agroecological systems due to their adaptive capabilities and potential contributions to sustainable agriculture. These breeds have evolved over generations to adapt to local conditions, developing natural defenses against prevalent pathogens and the ability to thrive in harsh environments (Edwards et al. 1993; Gill et al. 2014). Their efficient utilization of low-quality forage and grazing patterns that help maintain vegetation structure and control weed growth further underscore their importance in resource-efficient and ecologically sound livestock production (Niderkorn and Jayanegara 2021).

Conserving genetic diversity in livestock is not only essential for the resilience of agroecosystems but also plays a vital role in preserving cultural heritage and the social fabric of farming communities. Diverse livestock breeds are intrinsically linked to traditional farming practices and the traditional ecological knowledge (TEK) held by indigenous peoples and small-scale farmers. Protecting the intellectual property rights of these knowledge holders is crucial to ensure the continued preservation and sustainable use of these valuable genetic resources.

Overall, the conservation of diverse livestock breeds is a critical component of agroecological approaches, contributing to genetic diversity, resilience, and the preservation of cultural heritage.

7.2.3 The Multifaceted Benefits of Agroforestry Systems

Agroecology, a holistic approach to sustainable agriculture, advocates for the integration of trees and other perennial plants within agricultural landscapes, a practice known as agroforestry. Agroforestry systems offer a myriad of benefits, including enhanced biodiversity, improved soil health, optimized microclimates, and the creation of diverse habitats for a wide array of organisms.

The incorporation of diverse tree species in agroecological practices contributes significantly to on-farm biodiversity conservation. Agroforestry systems provide habitats for a variety of organisms, including birds, insects, and mammals, offering nesting sites, food sources, and shelter, thereby preserving and supporting diverse wildlife populations within agroecosystems (Pinho et al. 2012; Nair 2014). Furthermore, the presence of trees in agroforestry systems fosters improved soil health. Tree roots penetrate deeper into the soil profile, accessing nutrients and water that may be unavailable to shallow-rooted crops, which helps in nutrient cycling and prevents soil erosion by anchoring the soil and reducing runoff (Nair 2007, 2014; Pinho et al. 2012). Trees also contribute to the accumulation of organic matter through the deposition of leaves and other plant materials, enriching the soil and enhancing its fertility (Nair 2014).

7.2.4 Enhancing Agroecosystem Biodiversity: The Role of Habitat Creation and Conservation

Agroecology emphasizes the creation and preservation of diverse habitats within agricultural landscapes to support ecosystem services and biological pest control (Batáry et al. 2011). Hedgerows, buffer strips, ponds, and field margins are examples of features that provide refuge and food sources for beneficial insects, pollinators, and other wildlife. By maintaining and restoring these diverse habitats, agroecological farming systems can enhance biodiversity and promote the presence of natural enemies of pests (Altieri 1999; Batáry et al. 2011).

Hedgerows, linear rows of trees, shrubs, or perennial plants, act as ecological corridors that connect fragmented habitats and provide shelter and food resources for a wide range of organisms (Altieri 1999). These features are known to enhance biodiversity and promote the presence of natural enemies of pests, such as predatory insects and birds. Hedgerows also serve as important nesting sites for pollinators, including bees and butterflies, supporting their populations and enhancing pollination services in agricultural landscapes.

Buffer strips, areas of vegetation planted alongside water bodies or between agricultural fields, mitigate the movement of sediments, nutrients, and agrochemicals into waterways. These strips not only help protect water quality but also provide valuable habitat for beneficial insects, birds, and other wildlife (Power 2010; Forister et al. 2019). The presence of diverse plant species in buffer strips supports a range of beneficial organisms, including natural enemies of crop pests, thereby contributing to biological pest control (Batáry et al. 2011; Power 2010).

Ponds and wetlands within agricultural landscapes play a crucial role in supporting biodiversity and providing habitat for various aquatic and semi-aquatic species. They serve as breeding grounds for amphibians, provide water sources for wildlife, and support a diverse array of plants and invertebrates. These habitats enhance ecological balance and contribute to pest regulation through the presence of insect-eating organisms (Altieri 1999; Batáry et al. 2011; Power 2010; Forister et al. 2019).

A wetland established within an ecological farm

The preservation and restoration of diverse habitats in agroecological systems can provide a range of ecosystem services, including biological pest control and pollination, that can help mitigate the impacts of climate change on agricultural production (Power 2010).

7.2.5 Conserving Genetic Resources for Sustainable Agriculture: The Role of Agroecology

Agroecology recognizes the crucial importance of preserving genetic diversity for the future resilience and adaptability of agricultural systems (Francis et al. 2003). This includes safeguarding traditional crop varieties, landraces, and wild relatives, which often possess valuable traits such as drought tolerance, disease resistance, and enhanced nutritional quality (Thrupp 2000; Gill et al. 2014; Carlile and Garnett 2021). On-farm seed saving, participatory plant breeding, and community seed banks are practical approaches that contribute to the conservation of this genetic diversity and empower farmers in the process.

On-farm seed saving allows farmers to maintain and exchange diverse seed stocks, including locally adapted varieties, thereby reducing their dependence on external seed sources and giving them greater control over their seed supply (Thrupp 2000; Gill et al. 2014). Participatory plant breeding, which involves the collaboration of farmers, scientists, and breeders, enhances farmers' capacity to adapt their crops to specific agroecological conditions while also contributing to the preservation of diverse genetic resources. Community seed banks, where local communities collectively manage and preserve seeds of various crop varieties, further support the conservation of this invaluable biodiversity (Thrupp 2000; Gill et al. 2014; Ferede and Taye 2019; Flint-Garcia et al. 2023).

A well curated seed bank managed by the community

Preserving traditional crop varieties and landraces is crucial for maintaining genetic diversity and ensuring the resilience of agricultural systems in the face of changing environmental conditions (Gill et al. 2014; Flint-Garcia et al. 2023). These crop types often possess unique traits that enable them to thrive in specific local conditions, making them invaluable for adapting to the challenges posed by climate change (Thrupp 2000; Flint-Garcia et al. 2023). Furthermore, the protection of intellectual property rights for those with knowledge of the values and uses of such biodiversity, particularly for indigenous peoples and small farmers, is vital to incentivize their continued stewardship of these genetic resources (Thrupp 2000).

7.3 Nutrient Cycling in Agroecological Systems

Nutrient cycling is a fundamental concept in agroecology, as it underpins the sustainable and regenerative management of agricultural systems. Agroecology recognizes the importance of efficient nutrient cycling to maintain soil fertility, minimize nutrient losses, and reduce reliance on external inputs (Gliessman 1995).

7.3.1 Organic Matter Management

Agroecological practices prioritize the incorporation of organic matter into the soil, such as crop residues, cover crops, and animal manure. Organic matter decomposition contributes to the release of essential nutrients, including nitrogen, phosphorus, and potassium, as well as micronutrients, which become available to plants (Nayak et al. 2020). Unlike synthetic fertilizers that provide specific nutrients in a soluble form, organic matter releases nutrients slowly and in a more balanced manner, reducing the risk of nutrient imbalances and leaching.

The incorporation of organic matter into the soil also improves soil structure by enhancing soil aggregation and stability, which promotes better water infiltration, drainage, and root development. Furthermore, organic matter increases the soil's water-holding capacity, allowing it to retain moisture during dry periods and reducing the need for irrigation (Edwards et al. 1993).

7.3.2 Nutrient Cycling and Biodiversity

The maintenance of biological diversity and nutrient cycling mechanisms are global principles that are common to all agroecosystems and are essential in the design of sustainable agricultural systems. Agroecological practices that promote the use of cover crops, crop rotations, and diverse cropping systems contribute to the accumulation of organic matter, enhance nutrient cycling, and support a healthy and diverse soil biota.

These practices are informed by an understanding of ecosystem dynamics, where the complex set of biological, physical, and chemical interactions determine the processes that sustain agricultural productivity. By emulating the principles of natural ecosystems, agroecology seeks to create agricultural systems that are self-sustaining, stable, and able to maintain productivity with minimal external inputs (Gliessman 1995).

7.3.3 Crop Rotation and Diversification: Enhancing Agroecological Resilience

Agroecological systems often incorporate the practices of crop rotation and diversification, where different crop species are grown in sequence or intercropped. Crop rotation is a fundamental strategy in agroecology that involves cultivating a variety of crops in a planned sequence over time. This approach disrupts the life cycles of pests and diseases, as it reduces the availability of their host plants and limits the buildup of resistant populations.

In addition to pest and disease management, crop rotation improves soil health by enhancing nutrient cycling and organic matter accumulation. Different crop species have varying nutrient requirements and uptake patterns, allowing them to utilize different pools of nutrients in the soil (Smith et al. 2023). For example, leguminous crops can fix atmospheric nitrogen through symbiotic relationships with nitrogen-fixing bacteria, enriching the soil with this essential nutrient (Zhao et al. 2022). Furthermore, crop residues from different crops contribute to the organic matter content of the soil, improving soil structure and fertility when they decompose.

Crop diversification can also promote arthropod pest management by providing alternative food sources and habitats for natural enemies, as well as reducing the prevalence of host plants for pest species (Smith et al. 2023). A meta-analysis found that positive effects of crop rotation on soil aggregation and organic carbon are primarily influenced by climate and initial soil carbon content (Ball et al. 2005). Additionally, studies have shown that increasing the diversity of crop rotations can enhance cereal yields, providing economic and environmental benefits (Smith et al. 2023).

7.3.4 The Role of Green Manure and Cover Crops in Agroecological Practices

Agroecology promotes the use of green manure and cover crops as an integral part of practices aimed at improving soil fertility and reducing the reliance on external inputs (Norris and Congreves 2018). These crops are specifically grown and then incorporated into the soil, adding organic matter and releasing nutrients as they decompose (Andrew et al. 2012). Leguminous cover crops, for example, can fix atmospheric nitrogen and make it available to subsequent crops, reducing the need for synthetic nitrogen fertilizers (Sehgal et al. 2018).

The incorporation of green manure and cover crops offers multiple benefits for soil health and nutrient cycling (Liu et al. 2024). When these crops are grown, they capture carbon dioxide from the atmosphere through photosynthesis, incorporating it into their plant tissues (Andrew et al. 2012). The subsequent incorporation of these plants into the soil helps to increase soil organic matter content, leading to improved soil structure, water-holding capacity, and nutrient availability (Bolinder et al. 2020). The organic matter also serves as a food source for beneficial soil microorganisms, promoting their activity and enhancing overall soil health.

Leguminous cover crops have the additional advantage of nitrogen fixation. Through a symbiotic relationship with nitrogen-fixing bacteria, these crops can convert atmospheric nitrogen into a plant-usable form, thereby increasing soil nitrogen availability and reducing the need for synthetic nitrogen fertilizers. This helps to address the energy-intensive production and potential environmental pollution associated with overuse of synthetic nitrogen fertilizers.

The incorporation of green manure and cover crops can also help suppress weed growth by competing for resources and providing ground cover that shades out weed seedlings, reducing the need for herbicides (Bolinder et al. 2020; Nayak et al. 2020; Fontaine et al. 2024; Liu et al. 2024). The use of these practices aligns with the principles of agroecology, which aims to manage agroecosystems in a more sustainable and resilient manner (Andrew et al. 2012; Sehgal et al. 2018; Nayak et al. 2020; Fontaine et al. 2024).

7.3.5 Nutrient Cycling in Livestock-Integrated Agroecological Systems

Agroecology recognizes the critical role of livestock integration in agricultural systems for effective nutrient cycling. Livestock provide a valuable source of manure, which contains essential nutrients such as nitrogen, phosphorus, and potassium that are necessary for plant growth (Marschner and Rengel 2007). When properly managed and applied to cropland, these nutrients are gradually released and become available to plants over time, replenishing soil fertility and reducing the need for synthetic fertilizers (Marschner and Rengel 2007).

The incorporation of livestock manure into the soil not only supplies vital nutrients but also enhances the organic matter content, thereby improving soil structure, water-holding capacity, and nutrient retention. The organic matter in manure serves as a food source for beneficial soil microorganisms, stimulating their activity and promoting overall soil health.

However, to avoid negative environmental impacts, proper management of livestock waste is crucial. Techniques such as controlled application rates, timing, and incorporation into the soil help minimize nutrient losses through runoff and leaching, which can contribute to water pollution. Implementing practices that ensure nutrient application matches crop needs, and soil conditions can optimize nutrient utilization while minimizing environmental risks (Harmel et al. 2009; Nayak et al. 2020).

Agroecological systems also emphasize integrating livestock with crop production in ways that mimic natural ecosystems. For example, rotational grazing systems allow for rest and regrowth of vegetation while ensuring even nutrient distribution through manure deposition, creating a more sustainable and closed-loop nutrient cycle (Harmel et al. 2009; Verma et al. 2017; Nayak et al. 2020; Niderkorn and Jayanegara 2021).

7.3.6 *The Synergistic Benefits of Composting and Vermiculture in Agroecology*

Agroecology promotes the use of composting and vermiculture (earthworm-based composting) as methods to transform organic waste into nutrient-rich compost, which can then be applied to agricultural soils to enhance their fertility and productivity (Blouin et al. 2019; Nayak et al. 2020). Composting is a natural process that involves the decomposition of organic materials, such as crop residues, kitchen scraps, and animal manure, in the presence of oxygen. Microorganisms, including bacteria, fungi, and actinomycetes, break down the organic matter, converting it into stable humus-rich compost (Cooperband 2000). The application of compost to agricultural soils improves soil fertility and structure, increasing their water-holding capacity, nutrient retention, and cation exchange capacity.

Composting not only reduces the volume of organic waste but also creates a nutrient-rich and biologically active soil amendment (Nayak et al. 2020). Moreover, a systems perspective that includes full cost accounting of waste collection, handling, and processing must incorporate marketing, distributing, and recycling to make composting-based solid waste systems competitive.

Vermiculture, also known as worm composting, involves the use of earthworms to accelerate the decomposition of organic waste. Earthworms consume organic matter, partially digest it, and excrete it as nutrient-rich castings, also known as vermicompost. Vermicompost is highly beneficial to plant growth as it contains elevated levels of nutrients, plant growth regulators, and beneficial microorganisms. Vermiculture is a cost-effective tool for environmentally sound waste management, as earthworms aerate, condition, and fragment the substrate, thereby drastically altering the microbial activity and their biodegradation potential.

The synergistic effects of composting and vermiculture in agroecology are well-documented. Composting stabilizes organic matter, leading to long-lasting effects on soil fertility, while vermicomposting enhances nutrient availability through the action of earthworms. Integrating these two practices can provide a comprehensive approach to organic waste management and soil fertility enhancement, contributing to the overall sustainability of agroecological systems.

7.3.7 *Sustainable Nutrient Management in Agroecological Systems*

Agroecology emphasizes the importance of holistic nutrient management planning that considers the specific nutrient requirements of crops, soil conditions, and available nutrient sources. By carefully managing nutrient inputs and aligning them with the needs of individual crops and fields, agroecological practices aim to optimize nutrient use efficiency and minimize nutrient losses to the environment.

Efficient nutrient cycling is essential for maintaining soil fertility, reducing nutrient runoff, and promoting sustainable agricultural production in agroecological systems. Implementing practices that enhance nutrient cycling can minimize external inputs, reduce environmental impacts, and ensure long-term soil health and productivity. One key approach in agroecology is the use of organic amendments, such as compost and animal manure, which contribute to the overall nutrient pool in the soil (Fontaine et al. 2024). These organic sources release nutrients slowly over time, matching the uptake patterns of crops and reducing the risk of leaching or runoff (Edwards et al. 1993).

Another important agroecological practice is the use of cover crops and green manures, which add organic matter and nutrients to the soil. Legumes are particularly valuable as they fix atmospheric nitrogen and make it available to subsequent crops, reducing the need for synthetic fertilizers and enhancing nutrient cycling within the agroecosystem. Crop rotation is also widely used in agroecology to manage nutrient cycling effectively, as different crops have varying nutrient requirements and can access nutrients at different soil depths, reducing the risk of imbalances or depletion.

Overall, agroecological nutrient management strategies aim to create a closed-loop system that optimizes nutrient cycling, minimizes external inputs, and promotes long-term soil health and productivity (Edwards et al. 1993; Nayak et al. 2020; Swami 2020; Fontaine et al. 2024).

7.3.8 *Pest Regulation in Agroecological Systems: Harnessing Biodiversity and Ecological Principles*

Pest regulation is a crucial aspect of agroecology, an approach that emphasizes the use of ecological principles and practices to manage pests in agricultural systems. Agroecology recognizes the importance of minimizing reliance on synthetic pesticides and instead focuses on promoting natural pest regulation mechanisms.

One of the key elements of pest regulation in agroecology is the promotion of biodiversity and habitat diversity within agricultural landscapes. Diverse habitats provide refuge and food sources for beneficial organisms such as predatory insects, birds, bats, and spiders, which help regulate pest populations (Garibaldi et al. 2018).

By maintaining and enhancing biodiversity, agroecological systems foster natural pest control (Garibaldi et al. 2018).

Agroecology also promotes the use of conservation biological control, which involves the conservation and enhancement of natural enemies of pests (New 2005). This includes providing habitat, food sources, and alternative hosts for beneficial insects, such as parasitoids and predators. By creating favorable conditions for natural enemies, agroecological systems support the suppression of pest populations (Garibaldi et al. 2018).

Another key aspect of pest regulation in agroecology is the use of crop diversity and polyculture systems. Diverse cropping systems disrupt pest life cycles and reduce the buildup of pest populations. Different crop species may have varying susceptibilities to pests, reducing the risk of widespread pest outbreaks (Garibaldi et al. 2018). Polyculture systems can also confuse pests and provide a less favorable environment for their reproduction and survival (New 2005; Garibaldi et al. 2018; Forister et al. 2019).

Furthermore, agroecology emphasizes the use of cultural practices that help manage pests. This includes practices such as crop rotation, intercropping, trap cropping, and timing of planting to minimize pest pressure. Cultural practices can disrupt pest life cycles, create barriers to pest movement, and reduce the suitability of the agroecosystem for pest establishment and growth (New 2005).

7.3.9 Participatory Approaches in Agroecological Pest Management

Agroecology emphasizes farmer participation and knowledge sharing as integral components of effective pest management strategies, recognizing the invaluable insights and experiences that farmers possess regarding the dynamics of pests within their local ecosystems. Participatory approaches foster farmer empowerment, enabling them to actively engage in decision-making processes and contributing to the collective learning and adaptation of pest management practices.

One key aspect of this participatory approach is the use of Farmer Field Schools (FFS), which provide a platform for farmers to come together, observe pest problems, and collectively develop context-specific control measures. Through this collaborative process, farmers gain a deeper understanding of pest biology and ecology, leading to more effective and sustainable pest management solutions (Altieri 2002; Ali et al. 2017). Knowledge sharing is central to agroecological pest management, as farmers often possess valuable traditional practices and local insights that can be incorporated into these strategies.

Participatory approaches also empower farmers to make informed decisions regarding pest control based on their specific ecological and socioeconomic contexts, rather than relying solely on external experts. This empowerment is crucial for enhancing the effectiveness and sustainability of pest management practices, as farmers are able to tailor solutions to their unique circumstances. Furthermore, the principle of adaptive management, a fundamental tenet of agroecology, is facilitated by actively involving farmers in pest monitoring, evaluation, and decision-making (Altieri 2002; Ali et al. 2017). This allows for the identification of pest management practices that work best in specific local conditions, fostering a continuous process of learning and improvement (Altieri 2002; Ali et al. 2017; Forister et al. 2019).

7.3.10 The Role of Biological Control in Agroecological Systems

Agroecology, an approach that emphasizes the use of ecological principles and practices to manage pests in agricultural systems, recognizes the fundamental role of biological control as a key component in maintaining the balance and resilience of these complex ecosystems (Bale et al. 2008). Biological control involves the use of natural enemies, such as predators, parasitoids, and pathogens, to suppress pest populations and reduce crop damage, thereby promoting sustainable food production.

One of the core strategies in agroecology for enhancing natural pest control is the creation of habitat diversity. By incorporating diverse vegetation, such as hedgerows, flowering plants, and cover crops, agroecological systems provide shelter, nectar, and pollen sources for beneficial organisms, including predatory insects and parasitoids, which serve as natural enemies of agricultural pests (Harper 2001). These habitats act as refuges and nesting sites, thereby increasing the presence and activity of natural enemies within the agricultural landscape. Additionally, the preservation of natural habitats adjacent to farmland, such as field margins, woodlands, and wetlands, provides important reservoirs for beneficial organisms, ensuring their availability for pest control in agricultural fields (New 2005; Power 2010).

The emphasis on reducing pesticide use in agroecological practices is another key factor in promoting biological control. By minimizing the application of broad-spectrum pesticides, agroecology allows natural enemies to thrive, as these chemicals can often harm beneficial organisms that play a crucial role in regulating pest populations (New 2005; Bale et al. 2008; Lou et al. 2014; Campos et al. 2016).

The integration of biological control in agroecological systems can lead to significant economic, social, and environmental improvements. Biological control can be achieved through the introduction and management of natural enemies, such as parasitoids, predators, and entomopathogenic microorganisms, which can effectively suppress pest populations in a more sustainable manner than reliance on chemical pesticides (Bale et al. 2008; Lou et al. 2014; Campos et al. 2016). This approach not only reduces the need for pesticides, but also enhances biodiversity and the overall resilience of agricultural ecosystems, ultimately contributing to more sustainable food production.

7.3.11 Minimizing Pesticide Use in Agroecological Systems

Agroecology, an approach to sustainable agriculture, emphasizes the reduction of pesticide use, particularly broad-spectrum insecticides that can harm natural enemies. By minimizing pesticide applications, agroecological systems help preserve and support populations of beneficial organisms, ensuring their continued pest control services.

One of the key principles of agroecology is to prioritize non-chemical pest management strategies over the use of synthetic pesticides (Zanin et al. 2022; Soberón et al. 2023). This approach aims to create a balance between pest control and the conservation of beneficial organisms. Studies have shown that the reduction of pesticide use in agroecological systems positively impacts natural enemy populations, leading to effective pest control without compromising crop yields (Pretty and Pervez Bharucha 2015; Zanin et al. 2022; Soberón et al. 2023).

Agroecology promotes the implementation of alternative pest management strategies that minimize reliance on pesticides. These strategies include cultural practices such as crop rotation, intercropping, and the use of trap crops, as well as the promotion of biological control through the conservation of natural enemies. By adopting these practices, farmers can create an environment that supports natural enemies and reduces the need for chemical interventions.

In addition to the direct effects on natural enemies, the reduction of pesticide use in agroecological systems also has broader ecological benefits, helping prevent the negative impacts of pesticide drift and runoff on non-target organisms and the environment (Pretty and Pervez Bharucha 2015; Forister et al. 2019; Zanin et al. 2022; Soberón et al. 2023). As researchers have emphasized, society can support agricultural systems that move away from monocultures and heavy pesticide use, toward more diverse and ecologically based approaches that foster beneficial insects and minimize environmental harm (Pretty and Pervez Bharucha 2015; Forister et al. 2019).

7.3.12 Landscape-Level Approaches in Agroecology: Enhancing Biological Control Strategies

Agroecology recognizes the importance of considering the broader landscape in biological control strategies (New 2005). The design and management of agricultural landscapes can influence the abundance and effectiveness of natural enemies (Batáry et al. 2011; Francis et al. 2003). Agroecological practices aim to create landscape mosaics that provide suitable habitats for natural enemies, promote their movement, and enhance their ability to control pests across larger scales (Francis et al. 2003; New 2005; Bale et al. 2008; Batáry et al. 2011).

The arrangement and composition of habitats in the surrounding landscape can have a significant impact on the presence and activity of natural enemies in agricultural systems. Agroecology emphasizes the importance of incorporating diverse habitats within and around agricultural landscapes to support natural enemy populations. This includes the preservation of natural areas, such as forests, grasslands, and wetlands, as well as the creation of habitat features within agricultural fields, such as hedgerows, field margins, and flowering plants (Francis et al. 2003; Batáry et al. 2011).

Research has shown that landscape-level factors play a crucial role in shaping the abundance and effectiveness of natural enemies. Studies have demonstrated that the presence of natural habitats in the surrounding landscape enhances the population density and diversity of natural enemies in agricultural fields (Batáry et al. 2011; Martin et al. 2013). Similarly, the availability of floral resources in the landscape, through the presence of diverse flowering plants, can provide alternative food sources for natural enemies and enhance their survival and reproduction (New 2005; Batáry et al. 2011; Power 2010).

Agroecological practices aim to create landscape mosaics that promote connectivity between habitats and facilitate the movement of natural enemies. By providing corridors and stepping-stone habitats, agroecological landscapes support the dispersal of natural enemies, allowing them to colonize new areas and effectively suppress pests across larger scales (New 2005; Batáry et al. 2011; Power 2010; Martin et al. 2013). The cultural control practices employed in agroecology, such as habitat modifications and diversification, can contribute to the enhancement of biological control services provided by natural enemies (New 2005; Bale et al. 2008; Power 2010; Martin et al. 2013; Forister et al. 2019). These practices can create a more favorable environment for natural enemies, promoting their abundance and effectiveness in controlling agricultural pests (New 2005; Power 2010; Forister et al. 2019).

7.3.13 The Role of Agroecology in Greenhouse Gas Regulation and Mitigation

Agroecology plays a pivotal role in regulating and mitigating greenhouse gas (GHG). emissions through its emphasis on sustainable and climate-smart agricultural practices.

One of the key ways in which agroecology contributes to greenhouse gas regulation is through carbon sequestration. Agroecological practices focus on increasing soil organic matter, which results in the storage of carbon in agricultural soils (Leahy et al. 2020). The incorporation of organic residues, such as crop residues, animal manure, and compost, serves as a rich source of carbon and enhances soil organic matter. Studies have shown that the addition of organic amendments can increase soil carbon storage by up to 28% compared to conventional tillage practices (Favoino and Hogg 2008). Similarly, the adoption of cover cropping has been found to increase soil organic carbon by 15–20% compared to systems without cover crops (Morgan et al. 2010).

Agroforestry, the integration of trees and crops on the same land, is another agroecological practice that enhances carbon sequestration. Trees have the ability to capture and store large amounts of carbon in their biomass and in the soil through root growth and decomposition (Morgan et al. 2010; Liu et al. 2024).

In addition to carbon sequestration, agroecology also contributes to greenhouse gas regulation by reducing direct agricultural emissions. Climate-smart agricultural practices, such as optimizing fertilizer use, improving manure management, and adopting low-emission technologies, can help mitigate greenhouse gas emissions from agricultural production (Mesfin and Markos 2019).

7.3.14 Reducing Synthetic Fertilizer Use and Greenhouse Gas Emissions

Agroecology is an approach to agriculture that aims to minimize the use of synthetic fertilizers, particularly nitrogen fertilizers, due to their significant contribution to greenhouse gas (GHG). emissions (Barman et al. 2019). Synthetic nitrogen fertilizers are commonly used in conventional agriculture to meet the nutrient demands of crops, but their production, application, and subsequent reactions can result in the release of nitrous oxide (N_2O), a potent greenhouse gas with a significantly higher global warming potential than carbon dioxide (Rees et al. 2020; Fontaine et al. 2024). The Intergovernmental Panel on Climate Change (IPCC) estimates that nitrous oxide accounts for approximately 6% of total greenhouse gas emissions.

Agroecological practices focus on enhancing nutrient cycling within the farming system. By incorporating organic residues, adopting cover cropping, and practicing agroforestry, agroecological systems promote the recycling of nutrients, including nitrogen, within the agricultural ecosystem (Liu et al. 2024). The decomposition of organic matter releases nutrients slowly over time, reducing the need for synthetic nitrogen fertilizers. In addition to organic matter management, agroecological practices utilize natural fertilizers such as compost, animal manure, and legume cover crops. These natural fertilizers provide a more sustainable and environmentally friendly source of nutrients (Udvardi et al. 2015; Barman et al. 2019; Rees et al. 2020; Fontaine et al. 2024). Studies have shown the potential of agroecological practices to mitigate nitrous oxide emissions from agricultural soils through precision management (Rees et al. 2020) and the use of beneficial microorganisms in bio-fertilizers (Barman et al. 2019). The latest research suggests that these nature-based practices can improve long-term productivity of agrosystems while reducing the negative impact of agriculture on the environment and the climate (Udvardi et al. 2015; Barman et al. 2019; Fontaine et al. 2024).

7.3.15 Agroecological Approaches to Efficient Livestock Management: Mitigating Greenhouse Gas Emissions

Global livestock production is a significant contributor to greenhouse gas (GHG) emissions, primarily through methane (CH_4) and nitrous oxide (N_2O; Moran and Wall 2011). Agroecology recognizes the environmental challenges associated with livestock production and advocates for more sustainable and efficient livestock management practices to mitigate these emissions.

One such practice is rotational grazing, which involves moving livestock between different grazing areas to allow for pasture recovery and regeneration. This can optimize feed efficiency, as livestock have access to fresh, nutritious forage, reducing the need for supplemental feed. Additionally, rotational grazing can enhance nutrient cycling within the system, as livestock distribute their manure more evenly across the pasture, minimizing nutrient runoff and associated greenhouse gas emissions (Moran and Wall 2011; Moumen et al. 2016).

Silvopastoral systems, which integrate trees, forage crops, and livestock within the same land area, present another agroecological approach. Trees provide shade for livestock, improve microclimates, and contribute to carbon sequestration, while also enhancing nutrient cycling by capturing and recycling nutrients, reducing the need for external inputs. Studies have shown that silvopastoral systems can reduce greenhouse gas emissions compared to conventional livestock systems.

Mixed crop–livestock integration, which involves the simultaneous production of crops and livestock within the same farming system, is another strategy promoted by agroecology. Livestock can utilize crop residues and byproducts as feed, reducing the demand for external feed sources. By integrating crops and livestock, nutrient cycling is enhanced, as livestock manure becomes a valuable resource for crop fertilization, minimizing the need for synthetic fertilizers and contributing to the reduction of greenhouse gas emissions (Moran and Wall 2011; Shukla et al. 2019; Mo et al. 2023).

These agroecological practices offer promising solutions to the challenges of livestock production and greenhouse gas emissions. By optimizing feed efficiency, enhancing nutrient cycling, and integrating diverse elements within the farming system, these approaches can help mitigate the environmental impact of livestock production (Shukla et al. 2019; Niderkorn and Jayanegara 2021; Fontaine et al. 2024; Mo et al. 2023).

7.3.16 The Role of Agroforestry and Tree Planting in Sustainable Agriculture and Climate Change Mitigation

Agroecology recognizes the vital importance of integrating trees and agroforestry practices within agricultural systems, as they offer numerous environmental benefits and contribute to greenhouse gas (GHG) mitigation (Verchot et al. 2007; Lasco et al. 2014). Trees play a critical role in carbon sequestration, acting as long-term carbon sinks. Through the process of photosynthesis, trees absorb atmospheric carbon dioxide and store carbon in their biomass and soils. Agroforestry systems, which involve the deliberate integration of trees with crops or livestock, provide an opportunity for significant carbon sequestration, as the presence of trees in agroecological landscapes increases the overall carbon storage capacity of the system (Abbas et al. 2017).

An agroforest cover enhancing carbon sequestration

Agroforestry systems offer multiple benefits beyond carbon sequestration. They contribute to improved soil fertility by enhancing nutrient cycling and organic matter content, as the leaf litter and organic residues from trees provide a valuable source of nutrients for crops and promote soil health and productivity. Agroforestry also helps reduce soil erosion by providing effective windbreaks and reducing water runoff, thereby safeguarding soil resources and preventing sedimentation in water bodies (Nair 2007, 2014; Abbas et al. 2017; Koskey et al. 2021).

In addition to their role in carbon sequestration and soil health, trees in agroecological landscapes support enhanced biodiversity. The presence of diverse tree species creates habitat for a wide range of beneficial organisms, including insects, birds, and mammals, contributing to ecosystem resilience, pollination services, and natural pest control, reducing the reliance on synthetic pesticides.

Moreover, trees in agroecological landscapes can help reduce energy inputs and associated greenhouse gas emissions (Abbas et al. 2017). The shade provided by trees can mitigate heat stress in livestock and reduce the need for artificial cooling mechanisms, thus saving energy. While various agroforestry practices have been studied and implemented, the choice of tree species used in these systems depends on various environmental, social, and economic factors (Abbas et al. 2017; Koskey et al. 2021). As the benefits of agroforestry are increasingly recognized, modern land-use systems are evolving toward a more sustainable and holistic approach to land management (Nair 2007; Abbas et al. 2017).

References

Abbas F, Hammad HM, Fahad S, Cerdà A, Rizwan M, Farhad W, Ehsan S, Bakhat HF (2017, March 9) Agroforestry: a sustainable environmental practice for carbon sequestration under the climate change scenarios—a review. Environ Sci Pollut Res 24(12):11177–11191. https://doi.org/10.1007/s11356-017-8687-0

Ali MP, Bari MN, Ahmed N, Kabir M, Afrin S, Zaman M, Haque S, Willers JL (2017, May 1) Rice production without insecticide in smallholder farmer's field. Front Environ Sci 5:16. https://doi.org/10.3389/fenvs.2017.00016

Altieri MA (1999) The ecological role of biodiversity in agroecosystems. Agric Ecosyst Environ 74(1–3):19–31. https://doi.org/10.1016/s0167-8809(99)00028-6

Altieri MA (2002) Agroecology: the science of natural resource management for poor farmers in marginal environments. Agric Ecosyst Environ 93(1–3):1–24. https://doi.org/10.1016/s0167-8809(02)00085-3

Altieri MA, Nicholls CI (2013) The adaptation and mitigation potential of traditional agriculture in a changing climate. Clim Chang 140(1):33–45. https://doi.org/10.1007/s10584-013-0909-y

Andrew J, Kelton J, Mosjidis JA (2012, February 29) Utilization of Sunn Hemp for cover crops and weed control in temperate climates. Weed Control. https://doi.org/10.5772/36601

Bale JS, Van Lenteren JC, Bigler F (2008) Biological control and sustainable food production. Philos Trans R Soc B Biol Sci 363(1492):761–776. https://doi.org/10.1098/rstb.2007.2182

Ball BC, Bingham I, Rees RM, Watson CA, Litterick A (2005) The role of crop rotations in determining soil structure and crop growth conditions. Can J Soil Sci 85(5):557–577. https://doi.org/10.4141/s04-078

Barman S, Das S, Bhattacharya SS (2019) The prospects of bio-fertilizer technology for productive and sustainable agricultural growth. In: New and future developments in microbial biotechnology and bioengineering. Elsevier, pp 233–253. https://doi.org/10.1016/b978-0-444-64191-5.00017-1

Batáry P, Báldi A, Kleijn D, Tscharntke T (2011) Landscape-moderated biodiversity effects of agri-environmental management: a meta-analysis. Proc R Soc B Biol Sci 278(1713):1894–1902. https://doi.org/10.1098/rspb.2010.1923

Batish DR, Kohli RK, Jose S, Singh HP (2007) Ecological basis of agroforestry. CRC Press. https://doi.org/10.1201/9781420043365

Blouin M, Barrere J, Meyer N, Lartigue S, Barot S, Mathieu J (2019, June 17) Vermicompost significantly affects plant growth. A meta-analysis. Agron Sustain Dev 39(4):1–15. https://doi.org/10.1007/s13593-019-0579-x

Bolinder MA, Crotty F, Elsen A, Frac M, Kismányoky T, Lipiec J et al (2020) The effect of crop residues, cover crops, manures and nitrogen fertilization on soil organic carbon changes in agroecosystems: a synthesis of reviews. Mitig Adapt Strateg Glob Chang 25:929–952. https://doi.org/10.1007/s11027-020-09916-3

Campos EV, De Oliveira JL, Pascoli M, De Lima R, Fraceto LF (2016) Neem oil and crop protection: from now to the future. Front Plant Sci 7:216834. https://doi.org/10.3389/fpls.2016.01494

Carlile R, Garnett T (2021) What is agroecology. TABLE Explainer Series. TABLE, University of Oxford, Swedish University of Agricultural Sciences and Wageningen University & Research

Cooperband LR (2000) Composting: art and science of organic waste conversion to a valuable soil resource. Lab Med 31(5):283–290. https://doi.org/10.1309/w286-lqf1-r2m2-1wnt

Deutsch CA, Tewksbury JJ, Tigchelaar M, Battisti DS, Merrill SC, Huey RB, Naylor RL (2018) Increase in crop losses to insect pests in a warming climate. Science 361(6405):916–919. https://doi.org/10.1126/science.aat3466

Edwards CA, Grove TL, Harwood RR, Colfer CP (1993) The role of agroecology and integrated farming systems in agricultural sustainability. Agric Ecosyst Environ 46(1–4):99–121. https://doi.org/10.1016/0167-8809(93)90017-j

Erisman JW, van Eekeren N, de Wit J, Koopmans C, Cuijpers W, Oerlemans N, Koks BJ (2016) Agriculture and biodiversity: a better balance benefits both. AIMS Agric Food 1(2):157–174. https://doi.org/10.3934/agrfood.2016.2.157

Favoino E, Hogg D (2008) The potential role of compost in reducing greenhouse gases. Waste Manag Res 26(1):61–69. https://doi.org/10.1177/0734242x08088584

Ferede B, Taye W (2019) Sorghum genetic resource collection, evaluation, conservation and utilization in Ethiopia: a review. Int J Res 5(12):35–45. https://doi.org/10.20431/2454-6224.0512005

Flint-Garcia S, Feldmann MJ, Dempewolf H, Morrell PL, Ross-Ibarra J (2023) Diamonds in the not-so-rough: wild relative diversity hidden in crop genomes. PLoS Biol 21(7):e3002235. https://doi.org/10.1371/journal.pbio.3002235

Fontaine S, Abbadie L, Aubert M, Barot S, Bloor JM, Derrien D et al (2024) Plant–soil synchrony in nutrient cycles: learning from ecosystems to design sustainable agrosystems. Glob Chang Biol 30(1):e17034. https://doi.org/10.1111/gcb.17034

Forister ML, Pelton EM, Black SH (2019) Declines in insect abundance and diversity: we know enough to act now. Conserv Sci Pract 1(8):e80. https://doi.org/10.1111/csp2.80

Francis C, Lieblein G, Gliessman S, Breland TA, Creamer N, Harwood R et al (2003) Agroecology: the ecology of food systems. J Sustain Agric 22(3):99–118. https://doi.org/10.1300/j064v22n03_10

Garibaldi LA, Andersson GK, Requier F, Fijen TP, Hipólito J, Kleijn D, Rollin O (2018) Complementarity and synergisms among ecosystem services supporting crop yield. Glob Food Sec 17:38–47. https://doi.org/10.1016/j.gfs.2018.03.006

Gill BS, Raupp WJ, Friebe B (2014) Dual threats of imperiled native agroecosystems and climate change to world food security: genomic perspectives. J Crop Improv 28(1):88–98. https://doi.org/10.1080/15427528.2014.865411

Gliessman SR (1995) 3 Sustainable agriculture: an agroecological perspective. In: Advances in plant pathology, vol 11. Academic Press, pp 45–57. https://doi.org/10.1016/s0736-4539(06)80005-x

Gliessman SR (2020) Transforming food and agriculture systems with agroecology. Agric Hum Values 37:547–548. https://doi.org/10.1007/s10460-020-10058-0

Harmel RD, Smith DR, Haney RL, Dozier MONTY (2009) Nitrogen and phosphorus runoff from cropland and pasture fields fertilized with poultry litter. J Soil Water Conserv 64(6):400–412. https://doi.org/10.2489/jswc.64.6.400

Harper DR (2001) Biological control by microorganisms. e LS. https://onlinelibrary.wiley.com/doi/10.1038/npg.els.0004334

Jat SL, Suby SB, Parihar CM, Gambhir G, Kumar N, Rakshit S (2021) Microbiome for sustainable agriculture: a review with special reference to the corn production system. Arch Microbiol 203(6):2771–2793. https://doi.org/10.1007/s00203-021-02320-8

Kearns CA, Inouye DW, Waser NM (1998) Endangered mutualisms: the conservation of plant-pollinator interactions. Annu Rev Ecol Syst 29(1):83–112. https://doi.org/10.1146/annurev.ecolsys.29.1.83

Koskey G, Mburu SW, Awino R, Njeru EM, Maingi JM (2021) Potential use of beneficial microorganisms for soil amelioration, phytopathogen biocontrol, and sustainable crop production in smallholder agroecosystems. Front Sustain Food Syst 5:606308. https://doi.org/10.3389/fsufs.2021.606308

Lasco RD, Delfino RJP, Catacutan DC, Simelton ES, Wilson DM (2014) Climate risk adaptation by smallholder farmers: the roles of trees and agroforestry. Curr Opin Environ Sustain 6:83–88. https://doi.org/10.1016/j.cosust.2013.11.013

Leahy S, Clark H, Reisinger A (2020) Challenges and prospects for agricultural greenhouse gas mitigation pathways consistent with the Paris Agreement. Front Sustain Food Syst 4:69. https://doi.org/10.3389/fsufs.2020.00069

Lemly AD, Finger SE, Nelson MK (1993) Sources and impacts of irrigation drainwater contaminants in arid wetlands. Environ Toxicol Chem Int J 12(12):2265–2279. https://doi.org/10.1002/etc.5620121209

Liu L, Zhou W, Guan K, Peng B, Xu S, Tang J et al (2024) Knowledge-guided machine learning can improve carbon cycle quantification in agroecosystems. Nat Commun 15(1):357. https://doi.org/10.1038/s41467-023-43860-5

Lou YG, Zhang GR, Zhang WQ, Hu Y, Zhang J (2014) Reprint of: biological control of rice insect pests in China. Biol Control 68:103–116. https://doi.org/10.1016/j.biocontrol.2013.09.018

Marschner P, Rengel Z (2007, January 1) Nutrient cycling in terrestrial ecosystems. Soil Biol 10. https://doi.org/10.1007/978-3-540-68027-7

Martin EA, Reineking B, Seo B, Steffan-Dewenter I (2013) Natural enemy interactions constrain pest control in complex agricultural landscapes. Proc Natl Acad Sci 110(14):5534–5539. https://doi.org/10.1073/pnas.1215725110

Mesfin SG, Markos E (2019) A review on the role of climate smart agriculture for sustainable crop production. J Biol Agric Healthc 9:12. https://doi.org/10.7176/jbah/9-12-02

Mo L, Zohner CM, Reich PB, Liang J, De Miguel S, Nabuurs GJ et al (2023) Integrated global assessment of the natural forest carbon potential. Nature 624(7990):92–101. https://doi.org/10.1038/s41586-023-06723-z

Moran D, Wall E (2011) Livestock production and greenhouse gas emissions: defining the problem and specifying solutions. Anim Front 1(1):19–25. https://doi.org/10.2527/af.2011-0012

Morgan JA, Follett RF, Allen LH, Del Grosso S, Derner JD, Dijkstra F et al (2010) Carbon sequestration in agricultural lands of the United States. J Soil Water Conserv 65(1):6A–13A. https://doi.org/10.2489/jswc.65.1.6a

Moumen A, Azizi G, Chekroun KB, Baghour M (2016) The effects of livestock methane emission on the global warming: a review. Int J Global Warming 9(2):229–253. https://doi.org/10.1504/ijgw.2016.074956

Nair PR (2007) The coming of age of agroforestry. J Sci Food Agric 87(9):1613–1619. https://doi.org/10.1002/jsfa.2897

Nair PKR (2014) Agroforestry: practices and systems, an introduction to agroforestry, 270–282. https://doi.org/10.1016/b978-0-444-52512-3.00021-8

Nayak B, Rath BS, Shahid M, Jena SN, Bagchi TB, Roy PS (2020) Organic nutrient management in aromatic rice-rice sequence: a critical review. Int J Chem Stud 8:1435–1444. https://doi.org/10.22271/chemi.2020.v8.i5t.10503

New TR (2005) Cultural aspects of pest management. In: Invertebrate conservation and agricultural ecosystems, pp 189–212. https://doi.org/10.1017/cbo9780511542114.007

Niderkorn V, Jayanegara A (2021) Opportunities offered by plant bioactive compounds to improve silage quality, animal health and product quality for sustainable ruminant production: a review. Agronomy 11(1):86. https://doi.org/10.3390/agronomy11010086

Nodari RO, Guerra MP (2015) A agroecologia: estratégias de pesquisa e valores. Estudos avançados 29:183–207. https://doi.org/10.1590/s0103-40142015000100010

Norris CE, Congreves KA (2018) Alternative management practices improve soil health indices in intensive vegetable cropping systems: a review. Front Environ Sci 6:50. https://doi.org/10.3389/fenvs.2018.00050

Pinho RCD, Miller RP, Alfaia SS (2012) Agroforestry and the improvement of soil fertility: a view from amazonia. Hindawi Publ Corp Appl Environ Sci 2012:1–11. https://doi.org/10.1155/2012/616383

Potts SG, Biesmeijer JC, Kremen C, Neumann P, Schweiger O, Kunin WE (2010) Global pollinator declines: trends, impacts and drivers. Trends Ecol Evol 25(6):345–353. https://doi.org/10.1016/j.tree.2010.01.007

Power AG (2010) Ecosystem services and agriculture: tradeoffs and synergies. Philos Trans R Soc B Biol Sci 365(1554):2959–2971. https://doi.org/10.1098/rstb.2010.0143

Pretty J, Pervez Bharucha Z (2015) Integrated pest management for sustainable intensification of agriculture in Asia and Africa. Insects 6(1):152–182. https://doi.org/10.3390/insects6010152

Rai V, Sedeki AM, Parshad RD, Upadhyay RK, Bhowmick S (2011) Wetlands for water quality management – the science and technology. In: Current issues in water management, pp 163–176. https://doi.org/10.5772/27457

Rashid S, Mushtaq MB, Farooq I, Khan Z (2021) Climate smart crops for food security. In: The nature, causes, effects and mitigation of climate change on the environment. https://doi.org/10.5772/intechopen.99164

Rees RM, Maire JM, Florence A, Cowan N, Skiba UM, Van Der Weerden T, Ju X (2020) Mitigating nitrous oxide emissions from agricultural soils by precision management. Front Agric Sci Eng 7(1):75–80. https://doi.org/10.15302/j-fase-2019294

Sehgal M, Malik M, Singh RV, Kanojia AK, Singode A (2018) Integrated pest management in rice and its future scope. Int J Curr Microbiol App Sci 7(6):2504–2511. https://doi.org/10.20546/ijcmas.2018.706.297

Sethuraman G, Mohd Zain NA, Yusoff S, Ng YM, Baisakh N, Cheng A (2021) Revamping ecosystem services through agroecology—the case of cereals. Agriculture 11(3):204. https://doi.org/10.3390/agriculture11030204

Shukla PR, Skeg J, Buendia EC, Masson-Delmotte V, Pörtner HO, Roberts DC … Malley J (2019) Climate Change and Land: an IPCC special report on climate change, desertification, land degradation, sustainable land management, food security, and greenhouse gas fluxes in terrestrial ecosystems. https://philpapers.org/rec/SHUCCA-2

Smith ME, Vico G, Costa A, Bowles T, Gaudin AC, Hallin S et al (2023) Increasing crop rotational diversity can enhance cereal yields. Commun Earth Environ 4(1):89. https://doi.org/10.1038/s43247-023-00746-0

Soberón M, Bravo A, Blanco CA (2023, January 1) Strategies to reduce insecticide use in agricultural production. Elsevier eBooks, pp 507–514. https://doi.org/10.1016/b978-0-12-823960-5.00077-9

Stout JC, Dicks LV (2022) From science to society: implementing effective strategies to improve wild pollinator health. Philos Trans R Soc B 377(1853):20210165. https://doi.org/10.1098/rstb.2021.0165

Swami S (2020) Soil health management under organic production system. IJCS 8(2):330–339. https://doi.org/10.22271/chemi.2020.v8.i2e.8789

Thrupp LA (2000) Linking agricultural biodiversity and food security: the valuable role of agrobiodiversity for sustainable agriculture. Int Aff 76(2):265–281. https://doi.org/10.1111/1468-2346.00133

Udvardi M, Brodie EL, Riley W, Kaeppler S, Lynch J (2015) Impacts of agricultural nitrogen on the environment and strategies to reduce these impacts. Proc Environ Sci 29:303. https://doi.org/10.1016/j.proenv.2015.07.275

Verchot LV, Van Noordwijk M, Kandji S, Tomich T, Ong C, Albrecht A et al (2007) Climate change: linking adaptation and mitigation through agroforestry. Mitig Adapt Strateg Glob Chang 12:901–918. https://doi.org/10.1007/s11027-007-9105-6

Verma S, Singh A, Pradhan SS, Singh RK, Singh JP (2017) Bio-efficacy of organic formulations on crop production – a review. Int J Curr Microbiol App Sci 6(5):648–665. https://doi.org/10.20546/ijcmas.2017.605.075

Zanin ARA, Neves DC, Teodoro LPR, da Silva Júnior CA, da Silva SP, Teodoro PE, Baio FHR (2022) Reduction of pesticide application via real-time precision spraying. Sci Rep 12(1):5638. https://doi.org/10.1038/s41598-022-09607-w

Zhao J, Chen J, Beillouin D, Lambers H, Yang Y, Smith P et al (2022) Global systematic review with meta-analysis reveals yield advantage of legume-based rotations and its drivers. Nat Commun 13(1):4926. https://doi.org/10.1038/s41467-022-32464-0
Zhu JQ, Yu H, He QY (2013, January) Development of wetland agriculture and its prospect. In: 2013 Third international conference on intelligent system design and engineering applications. IEEE, pp 596–601. https://doi.org/10.1109/isdea.2012.145

Chapter 8
Ecology-Based Concepts of Sustainable Agriculture

Ecology-based concepts form the foundation of sustainable agriculture, as they emphasize the integration of ecological principles and processes into agricultural systems.

8.1 Biodiversity Conservation in Sustainable Agriculture

Sustainable agriculture recognizes the vital role of biodiversity in maintaining ecosystem health and resilience. It aims to conserve and enhance biodiversity within agricultural landscapes through diverse practices such as crop rotations, intercropping, agroforestry, and the preservation of natural habitats (Altieri 1999; Cozim-Melges et al. 2024). This biodiversity supports key ecosystem services like natural pest control, nutrient cycling, soil fertility, and overall ecosystem stability (Altieri 1999).

Sustainable agriculture prioritizes the maintenance and improvement of soil health, emphasizing techniques that enhance soil organic matter, promote soil structure, and minimize soil erosion. Cover cropping, composting, reduced tillage, and crop diversification are some of the practices that help build and protect the soil's fertility, structure, and nutrient cycling capacity.

Sustainable agriculture also focuses on efficient nutrient cycling within agroecosystems. It aims to reduce external inputs of synthetic fertilizers and instead relies on organic matter decomposition, legume nitrogen fixation, and nutrient recycling through diverse cropping systems and livestock integration. By optimizing nutrient cycling, sustainable agriculture minimizes nutrient loss to the environment, reduces pollution, and improves resource-use efficiency (Altieri 1999; Thrupp 2000; Cozim-Melges et al. 2024).

M. Otieno, *Sustainable Agroecological Practices in Sub-Saharan Africa in the Face of Climate Change*, Advances in Geographical and Environmental Sciences, https://doi.org/10.1007/978-3-031-70472-7_8

Water management is another crucial aspect of sustainable agriculture. It includes techniques such as precision irrigation, rainwater harvesting, water-efficient cropping systems, and water conservation measures. By minimizing water use, optimizing irrigation, and protecting water quality, sustainable agriculture ensures the efficient and sustainable use of this vital resource.

Sustainable agriculture further adopts an integrated approach to pest management, emphasizing prevention, monitoring, and the use of multiple strategies to manage pests effectively. Integrated Pest Management (IPM) prioritizes cultural, biological, and mechanical control methods over the heavy use of synthetic pesticides (Batáry et al. 2011; Cozim-Melges et al. 2024).

8.2 Conservation Agriculture: Principles, Practices, and Relationship with Agroecology

Conservation agriculture (CA) and agroecology are two interconnected approaches that share similar principles and goals in promoting sustainable and environmentally friendly agricultural practices (Francis et al. 2003; Gliessman 2020). Conservation agriculture (CA) is an approach that emphasizes sustainable farming methods, including reduced or no-tillage, to minimize soil disturbance and maintain soil health. Agroecology, on the other hand, is a holistic approach that seeks to reconcile agriculture and local communities with natural processes for the common benefit of nature and livelihoods (Suárez-Torres et al. 2017).

Conservation agriculture is based on three main principles: minimal soil disturbance, preservation of crop residues, and crop rotation. Reduced or no-tillage practices play a crucial role in conservation agriculture by preserving soil structure and reducing erosion. These practices can enhance soil aggregation, improve water infiltration and retention, and contribute to better root growth and nutrient availability for plants. The key tenet of conservation agriculture is to utilize agricultural biomass effectively while avoiding biomass degradation and the associated production of greenhouse gases.

The relationship between conservation agriculture and agroecology is complementary, as both approaches share the common goal of promoting sustainable agriculture. Conservation agriculture focuses on specific practices, such as reduced tillage and crop rotation, to maintain soil health and productivity, while agroecology takes a more holistic view, considering the broader interactions within the agricultural system (Gliessman 1995, 2020; Suárez-Torres et al. 2017; Sethuraman et al. 2021). By integrating the principles of conservation agriculture and the systems-level approach of agroecology, farmers and researchers can develop more resilient and sustainable agricultural systems that benefit both the environment and local communities.

8.3 Integrating Ecological Principles into Agricultural Systems

Agroecological practices have emerged as a comprehensive approach to sustainable agriculture that prioritizes the integration of ecological principles into agricultural systems (Francis et al. 2003). This paradigm shift from conventional, intensive farming methods to more ecologically sound techniques has the potential to enhance biodiversity, build resilience, and provide a wider range of ecosystem services.

An intensively managed agricultural landscape

The 10 elements of agroecology provide a robust framework for understanding this multifaceted concept (Barrios et al. 2020). At its core, agroecology emphasizes the importance of adapting farming practices to the unique characteristics of a given place, recognizing the inherent value and interconnectedness of the natural world (Francis et al. 2003). By mimicking the complexity and diversity of natural ecosystems, agroecological approaches can enhance the provision of vital services such as pollination, nutrient cycling, and biological pest control (Sethuraman et al. 2021).

Agroecological transitions are particularly relevant for staple crops like cereals, which are essential for global food security. Diversifying crop rotations, integrating livestock, and minimizing synthetic inputs can help revitalize ecosystem services and improve the resilience of cereal-based farming systems in the face of climate change. Beyond the farm gate, agroecology also emphasizes social and economic dimensions of sustainability, including strengthening local food systems and empowering rural communities (Sethuraman et al. 2021; Cozim-Melges et al. 2024).

Research has demonstrated the efficacy of agroecological practices in enhancing biodiversity across diverse biomes. As the global community confronts the pressing challenges of food system transformation, agroecology offers a holistic pathway to reconcile economic, environmental, and social goals.

8.3.1 The Importance of Crop Diversity in Agroecological Farming

Agroecology, an agricultural approach that emphasizes the significance of diverse cropping systems, offers numerous benefits ranging from enhanced biodiversity to reduced pest and disease pressure and improved soil fertility (Gliessman 2020; Sethuraman et al. 2021).

One of the key advantages of diverse cropping systems is the promotion of biodiversity. By cultivating a variety of crops, farmers create a habitat that attracts a wide range of beneficial organisms, such as pollinators and natural predators of pests, contributing to the overall ecological balance of the agroecosystem and reducing the reliance on chemical inputs for pest control (Barzman et al. 2015). Furthermore, crop diversity helps in reducing pest and disease pressure.

Monocultures, which involve the cultivation of a single crop over a large area, are more susceptible to pest outbreaks and disease epidemics. In contrast, diverse cropping systems disrupt pest life cycles and create a less favorable environment for pests to establish and spread. Different crops may possess varying levels of resistance or tolerance to specific pests and diseases, effectively acting as natural barriers against their proliferation.

In addition to biodiversity and pest management benefits, diverse cropping systems contribute to improved soil fertility. Different crops have varying nutrient requirements, root structures, and growth patterns, ensuring a more efficient utilization of soil nutrients and preventing excessive depletion of specific nutrients (Gliessman 2020). Additionally, certain crops, such as legumes, have the ability to fix atmospheric nitrogen, further enhancing soil fertility and reducing the need for synthetic fertilizers (Gliessman 2020).

Agroecology, with its emphasis on crop diversity, has been recognized as a stepping stone to achieving several Sustainable Development Goals, as it has the potential to build climate change-resilient farming systems while enhancing ecosystem services and reducing biodiversity loss (Sethuraman et al. 2021).

Crop diversity in a smallholder farmer's establishment

8.3.2 *The Multifaceted Benefits of Agroforestry*

Agroforestry, an agricultural practice that strategically integrates trees or shrubs with crops or livestock, has emerged as a sustainable approach to land management, offering a range of advantages that extend beyond the realm of food production. One of the primary benefits of agroforestry is its positive impact on soil fertility. The presence of trees in agroforestry systems facilitates nutrient cycling and organic matter accumulation, resulting in improved soil structure and fertility (Nair 2007, 2013, 2014). Through processes like nitrogen fixation and nutrient uptake, trees contribute to the nutrient enrichment of the soil, thereby enhancing the growth and productivity of associated crops.

Microclimate regulation is another key advantage of agroforestry. The strategic placement of trees in agroforestry systems helps moderate temperature extremes, reduce wind speed, and provide shade to crops and livestock. These microclimatic modifications create a more favorable environment for plant growth and animal welfare, reducing heat stress and enhancing overall productivity.

Additionally, agroforestry systems promote enhanced biodiversity. The combination of trees, crops, and livestock provides a variety of habitats, food sources, and nesting sites, attracting diverse flora and fauna. The presence of trees in agroforestry landscapes supports beneficial insects, birds, and other wildlife, which can contribute to natural pest control and pollination services.

Furthermore, agroforestry offers additional sources of income for farmers. Trees in agroforestry systems can provide timber, fruits, nuts, medicinal products, and other valuable non-timber forest products, generating a diverse array of revenue streams for the landowner (Pinho et al. 2012; Nair 2014).

As the plethora of benefits of agroforestry are realized, modern land-use systems are evolving toward a more sustainable and holistic approach to land management that integrates the strategic use of trees, crops, and livestock (Nair 2007, 2013, 2014).

8.3.3 The Benefits of Organic Farming: Promoting Soil Health, Biological Pest Control, and Sustainable Nutrient Management

Organic farming, a central component of agroecology, emphasizes the utilization of natural inputs and the avoidance of synthetic pesticides and fertilizers (Lee et al. 2019). This approach to agriculture has gained increasing attention due to its focus on soil health, biological pest control, and sustainable nutrient management (Dotaniya et al. 2020; Swami 2020).

One of the key principles of organic farming is prioritizing soil health (Lee et al. 2019; Swami 2020) Organic farmers employ techniques such as crop rotation, cover cropping, and the application of organic matter to build and maintain soil fertility (Dotaniya et al. 2020). By fostering healthy soils, organic farming supports the growth of robust plants, improves nutrient cycling, and enhances overall ecosystem resilience (Lee et al. 2019; Patel and Champaneri 2020; Swami 2020). Biological pest control is another fundamental aspect of organic farming. Instead of relying on synthetic pesticides, organic farmers promote the use of beneficial insects, natural predators, and cultural practices to manage pests.

This approach minimizes the need for chemical interventions while effectively controlling pest populations. Composting also plays a significant role in organic farming systems, as it is used to improve soil structure, fertility, and water-holding capacity. Composting enhances microbial activity in the soil, contributing to nutrient availability and supporting the growth of healthy crops (Lee et al. 2019; Patel and Champaneri 2020; Nayak et al. 2020; Swami 2020).

Organic nutrient management is an essential practice in organic farming. Organic farmers rely on natural sources of nutrients, such as compost, animal manures, and green manures, to supplement soil fertility and enhance crop production. Unlike conventional farming, organic agriculture aims to maintain a sustainable balance between inputs and outputs, minimizing the reliance on external, synthetic inputs (Lee et al. 2019; Patel and Champaneri 2020).

8.3.4 The Importance of Conservation Tillage Practices in Sustainable Agriculture

Conservation tillage practices play a crucial role in promoting sustainable agriculture by minimizing soil disturbance and maintaining soil cover (Zheng et al. 2014). These practices, which include reduced tillage, no-till, and strip-till techniques, aim to retain crop residues on the soil surface, improve water infiltration, and preserve soil structure (Tamburini et al. 2016; Arriaga et al. 2017).

Reduced tillage is a conservation tillage practice that involves minimizing the frequency and intensity of tillage operations compared to conventional tillage systems. By reducing the number of tillage passes, reduced tillage helps to maintain soil structure, organic matter content, and microbial activity, contributing to improved water infiltration, soil moisture retention, and nutrient cycling (Tamburini et al. 2016). No-till is a conservation tillage technique that eliminates or significantly reduces soil disturbance, with crops planted directly into untilled soil. The retention of crop residues protects the soil from erosion, enhances water infiltration, and reduces evaporation, while also promoting soil organic matter accumulation, which improves soil structure, nutrient availability, and water-holding capacity (Blevins and Frye 1993; Tamburini et al. 2016; Arriaga et al. 2017). Strip-till is another conservation tillage approach that combines elements of reduced tillage and no-till, where tillage is limited to narrow strips where seeds or transplants are placed, leaving the rest of the field untilled (Blevins and Frye 1993; Tamburini et al. 2016; Arriaga et al. 2017). Conservation tillage practices have been shown to provide a range of benefits, including improved soil health, enhanced water infiltration, and reduced erosion (Blevins and Frye 1993; Tamburini et al. 2016; Arriaga et al. 2017).

8.3.5 *Integrated Pest Management: A Comprehensive Approach to Sustainable Agriculture*

Integrated Pest Management (IPM) is an ecologically based approach to pest management that aims to minimize the reliance on synthetic pesticides while effectively managing pest populations (SourceWatch 2008; Juroszek and Von Tiedemann 2011). This method involves the integration of various strategies, including cultural practices, biological control, targeted pesticide use, monitoring, and the promotion of natural enemies, to create a more sustainable and integrated approach to pest management.

Cultural practices play a significant role in IPM (Pretty and Pervez Bharucha 2015). These practices involve modifying the crop environment to create conditions that are less favorable for pests and more favorable for beneficial organisms (Pretty and Pervez Bharucha 2015). Examples of cultural practices include crop sanitation, proper irrigation and drainage, appropriate fertilization, and crop rotation (Juroszek and Von Tiedemann 2011). These practices help reduce pest pressure by minimizing pest habitat, enhancing crop health and resilience, and disrupting pest life cycles. Biological control is an essential component of IPM, involving the use of natural enemies, such as predators, parasitoids, and pathogens, to regulate pest populations. By promoting the presence and activity of these beneficial organisms, either through conservation or augmentation strategies, biological control can provide effective and sustainable pest management (Juroszek and Von Tiedemann 2011; Pretty and Pervez Bharucha 2015). Targeted pesticide use is another aspect of IPM, but it is used judiciously and only when necessary. When pest populations exceed economic thresholds or when cultural and biological control measures are insufficient, targeted pesticide applications may be employed. The selection of pesticides is based on careful consideration of their efficacy, selectivity, and potential impact on the environment and non-target organisms (Juroszek and Von Tiedemann 2011; Pretty and Pervez Bharucha 2015).

The integrated approach of IPM offers a more sustainable and effective way to manage pests, reducing reliance on synthetic pesticides and maintaining the natural balance of ecosystems. This approach not only benefits the environment but also reduces costs for farmers and ensures the long-term viability of agricultural production.

8.4 Water Management in Agroecological Systems

Agroecology, as a sustainable agricultural approach, prioritizes water-efficient practices to optimize water use and enhance water resource management. These practices aim to minimize water waste, mitigate the risk of water-related diseases, and improve overall water management in agricultural systems.

Drip irrigation is a water-efficient technique commonly advocated in agroecology. It involves the slow and precise application of water directly to the plant's root zone through a network of tubes or emitters, ensuring that water is delivered where it is needed most and reducing evaporation and runoff compared to traditional irrigation methods (Perakis et al. 2017; Reddy et al. 2017; El-Shafei and Mattar 2022). This practice optimizes water-use efficiency, as it reduces water loss through evaporation and minimizes weed growth by delivering water only to the targeted plants (El-Shafei and Mattar 2022).

Rainwater harvesting is another water-efficient practice promoted by agroecology. It involves capturing and storing rainwater for agricultural use during periods of water scarcity. Rainwater harvesting systems can include various components such as rooftop catchments, storage tanks, and distribution networks. By harnessing rainwater, agroecological systems can supplement irrigation needs, particularly during dry spells, and reduce dependence on other water sources. This practice not only conserves water but also helps replenish groundwater resources and enhances resilience to drought conditions.

Proper water scheduling is a crucial aspect of agroecology that ensures efficient water use. It involves timing irrigation applications based on crop water requirements, soil moisture levels, and environmental conditions to optimize water use and minimize waste. Deficit irrigation, a water management technique that allows larger agricultural lands to be irrigated with scarce water resources, can also be integrated into agroecological systems by carefully considering the timing and degree of water stress to which plants are exposed (Patanè et al. 2011; He et al. 2018; El-Shafei and Mattar 2022). The optimization of water-land combinations and planting patterns can further contribute to water-efficient agroecological practices (Asrat 2018).

8.5 Agroecological Practices for Sustainable Soil Conservation

Agroecology, a holistic approach to agricultural systems, places a strong emphasis on soil conservation, recognizing the vital role of healthy soils in sustainable land management. This research paper will elaborate on the key soil conservation techniques promoted by agroecology, including contour plowing, terracing, and the use of cover crops, all of which contribute to long-term sustainable land management.

Contour plowing, a soil conservation technique commonly advocated in agroecology, involves plowing along the contour lines of the land, following the natural slope of the terrain. By plowing perpendicular to the slope, contour plowing helps to create ridges and furrows that act as barriers, reducing the flow of water and minimizing soil erosion. This technique effectively slows down the movement of water across the field, allowing more time for water infiltration and reducing the risk of soil runoff.

Terracing is another soil conservation practice promoted by agroecology, especially in hilly or sloping landscapes. Terraces are horizontal or nearly horizontal platforms constructed on the contours of the land, designed to slow down water flow, prevent soil erosion, and retain water within each terrace. By creating a series of level platforms, terracing helps to break the slope of the land, reducing the velocity of water runoff and facilitating water infiltration into the soil. This technique not only helps prevent soil erosion but also conserves soil moisture, making it especially beneficial in regions with limited rainfall or in areas prone to drought.

The use of cover crops is another key strategy in agroecological soil conservation. Cover crops are specific plant species that are grown primarily to protect and enrich the soil rather than for harvest (Oyana et al. 2015; Bolinder et al. 2020; Sanam et al. 2022). These plants help to prevent soil erosion, retain soil moisture, and improve soil fertility through their root systems and organic matter contributions.

8.6 Livestock Integration in Agroecological Systems

Agroecology, as a comprehensive approach to agriculture, emphasizes the integration of livestock into agricultural systems to enhance nutrient cycling and diversify farm enterprises. This paper will elaborate on the practices promoted by agroecology for livestock integration, including rotational grazing, mixed crop-livestock systems, and the use of animal manure as organic fertilizer. These practices aim to improve soil fertility, reduce pests, and increase overall farm productivity (Francis et al. 2003).

Rotational grazing is a key practice advocated in agroecology for sustainable livestock management (Cozim-Melges et al. 2024). It involves dividing pasture into smaller paddocks and systematically moving livestock from one paddock to another

(Cozim-Melges et al. 2024). This practice allows for adequate rest and regrowth of pasture plants, preventing overgrazing and promoting healthy plant growth (Cozim-Melges et al. 2024). By moving livestock frequently, rotational grazing also improves nutrient cycling as animal waste is more evenly distributed across the pasture, which helps improve soil fertility, reduce soil compaction, and enhance overall pasture health and productivity. Mixed crop-livestock systems are another important component of agroecology. These systems involve integrating livestock production with crop production on the same farm or land unit. The integration of crops and livestock allows for the recycling of nutrients and energy, creating a more sustainable and efficient agricultural system. For example, crop residues can be used as feed for livestock, and animal manure can be used as organic fertilizer for crops. This integration helps improve soil fertility through nutrient cycling and reduces the reliance on external inputs (Niderkorn and Jayanegara 2021; Fontaine et al. 2024; Cozim-Melges et al. 2024). Additionally, mixed crop-livestock systems provide economic diversification and increased resilience for farmers (Niderkorn and Jayanegara 2021; Fontaine et al. 2024; Cozim-Melges et al. 2024).

Dairy cows at an agroecological milk production establishment

The use of animal manure as organic fertilizer is another key practice in agroecological livestock integration. Animal manure can be a valuable source of nutrients for crops, replacing the need for synthetic fertilizers. When applied appropriately, manure can improve soil fertility, increase water-holding capacity, and enhance overall soil health (Teague 2018). Combining these practices of rotational grazing, mixed crop-livestock systems, and manure application can create a more closed-loop, integrated agricultural system that mimics natural ecosystems, reducing external inputs and enhancing the environmental and economic sustainability of farming operations.

8.7 The Role of Cover Crops in Sustainable Agriculture

Cover crops are a fundamental component of agroecological practices, playing a crucial role in maintaining soil health, enhancing ecosystem services, and promoting sustainable agricultural systems. These non-cash crops are grown primarily to cover and protect the soil between main crop rotations, providing a multitude of benefits that contribute to the long-term viability of farmlands.

One of the primary advantages of cover crops is their ability to improve soil structure and water-holding capacity, thereby increasing the effectiveness of applied nitrogen fertilizers (Lu et al. 2000). Legume cover crops, such as hairy vetch and crimson clover, can even fix atmospheric nitrogen and contribute to the nitrogen requirements of subsequent cash crops. Cover crops can also suppress weeds, provide habitat for beneficial predator insects, and act as non-host crops for nematodes and other pests in crop rotations. This not only reduces the need for chemical inputs, but also promotes the development of integrated pest management strategies.

Numerous studies have demonstrated the agronomic and economic benefits of incorporating cover crops into sustainable food production systems. Research has shown that the profitability of cover crops is often a result of enhanced crop yields rather than reduced input costs. By improving soil health, cover crops can increase the availability of nutrients and water, leading to higher yields and greater overall productivity (Outreach 2007).

In addition to their direct agricultural benefits, cover crops also contribute to the overall health and resilience of the agroecosystem. By covering the soil and preventing erosion, cover crops help to protect against the dislodging and movement of soil particles, which can be carried away by water or wind. This, in turn, helps to maintain the long-term viability of farmlands and reduce the risk of soil degradation (Outreach 2007; Andrew et al. 2012).

8.7.1 The Role of Cover Crops in Soil Erosion Prevention

Cover crops play a crucial role in agroecological practices by effectively preventing soil erosion through their ability to provide ground cover and reduce the impact of raindrops on bare soil (Kavian et al. 2020). The dense vegetation of cover crops acts as a physical barrier that shields the soil surface from the impact of raindrops, intercepting the raindrops and preventing their direct contact with the soil surface. This interception significantly reduces the kinetic energy of raindrops, minimizing their ability to dislodge soil particles and erode the soil surface (Kumawat et al. 2020).

Moreover, the dense vegetation and intricate root systems of cover crops help stabilize the soil structure and increase its resistance to erosion. The above-ground biomass of cover crops acts as a physical barrier, providing ground cover that reduces the velocity of water flow across the soil surface. By slowing down the movement of water, cover crops allow more time for water infiltration into the soil, reducing surface runoff and minimizing erosion. Furthermore, the root systems of cover crops play a vital role in preventing soil erosion. The extensive and fibrous root networks of cover crops bind the soil particles together, enhancing soil stability and preventing soil detachment. These roots act as anchors, holding the soil in place and reducing the likelihood of soil erosion (Andrew et al. 2012). Additionally, the root channels created by cover crop roots can improve water infiltration and reduce surface runoff, further contributing to erosion control (Outreach 2007; Andrew et al. 2012; Marioti et al. 2013; Kumawat et al. 2020). Cover crops contribute indirectly to overall soil health by catching nutrients before they can leach out of the soil profile or, in the case of legumes, by adding nitrogen to the soil. Their roots can even help unlock some nutrients, converting them to more available forms (Lu et al. 2000; Outreach 2007). Cover crops provide habitat or a food source for some important soil organisms, break up compacted layers in the soil, and help dry out wet soils. The effectiveness of cover crops in soil erosion prevention depends on various factors, including rainfall intensity, soil type, and topography of a particular locality. The selection of the right cover crop is crucial for soil and water conservation, and it should be based on the intensity and critical period of rainfall, market demand, climate, and resources of the farmer (Outreach 2007; Andrew et al. 2012; Gabriel et al. 2019; Kumawat et al. 2020).

8.7.2 Use of Legume Cover Crops for Sustainable Soil Fertility Improvement

Sustainable agroecological systems rely on the strategic use of cover crops, particularly legumes, to enhance soil fertility through biological nitrogen fixation. Legume cover crops, such as clover, vetch, and pea, form symbiotic relationships with rhizobia bacteria present in their root nodules. These bacteria have the remarkable ability to convert atmospheric nitrogen into ammonia, which is then converted into forms like nitrate that can be readily absorbed by plants. Through this symbiotic relationship, legume cover crops have the unique ability to enrich the soil with nitrogen, providing an essential nutrient that supports plant growth and development, without relying on synthetic fertilizers.

The nitrogen released by legume cover crops offers numerous benefits to agroecological systems. Firstly, it enriches the soil with nitrogen, a critical component of proteins, enzymes, and chlorophyll, all of which are vital for various plant functions. The availability of nitrogen from legume cover crops reduces the reliance on external nitrogen inputs, such as synthetic fertilizers, promoting sustainable agricultural practices. In addition to nitrogen fixation, cover crops contribute organic matter to the soil through their root systems and above-ground biomass, improving soil structure and water infiltration (Lu et al. 2000; Fageria et al. 2005; Venkateshwaran and Ané 2011).

Legumes have the ability to establish a mutualistic association with soil bacteria known as rhizobia, which form root nodules, inside of which atmospheric dinitrogen is reduced into assimilable forms and supplied to host plants. A high level of species specificity exists in rhizobia–legume interactions, which are initiated by a signal exchange between the two partners (Venkateshwaran and Ané 2011). Legume crops meet their nitrogen requirement through symbiotic nitrogen fixation, reducing the dependency of agricultural crops on fossil fuel-derived nitrogenous fertilizers. Additionally, biologically fixed nitrogen is bound in soil organic matter and thus much less susceptible to soil chemical transformations and physical factors that lead to volatilization and leaching (Lu et al. 2000; Venkateshwaran and Ané 2011; Mus et al. 2016).

Incorporating legume cover crops into agroecological systems offers a sustainable approach to improving soil fertility and reducing reliance on synthetic fertilizers. The nitrogen fixation capabilities of legumes, coupled with their ability to enhance soil structure and organic matter, make them a valuable asset in promoting long-term soil health and productivity (Lu et al. 2000; Venkateshwaran and Ané 2011; Mus et al. 2016).

8.7.3 Weed Suppression and Pest Management in Agroecological Systems

Cover crops play a crucial role in sustainable agriculture by providing various benefits, including weed suppression and pest management (Andrew et al. 2012; Sehgal et al. 2018). One of the primary mechanisms by which cover crops suppress weeds is through shading the soil surface, which limits the amount of light available for weed germination and growth (Lu et al. 2000). The dense vegetation of cover crops forms a canopy that effectively outcompetes weeds for essential resources such as light, water, and nutrients, inhibiting their establishment and development (Sehgal et al. 2018).

Furthermore, cover crops can physically smother weeds, preventing them from accessing necessary resources. The thick biomass and ground cover provided by cover crops create a physical barrier that hinders weed emergence and growth, reducing weed competitiveness and population size (Andrew et al. 2012).

In addition to competition for resources and physical suppression, some cover crops exhibit allelopathic properties, releasing natural compounds that can inhibit the growth of weeds and other pests. Certain cover crops produce allelochemicals, such as phenolic compounds, glucosinolates, or terpenoids, which can act as biopesticides, inhibiting weed germination, growth, and reproduction (Andrew et al. 2012; Sehgal et al. 2018).

This approach aligns with the principles of integrated pest management (IPM), which emphasizes the use of cultural, biological, and mechanical control methods to reduce reliance on synthetic pesticides (Lu et al. 2000; Andrew et al. 2012; Sehgal et al. 2018). The development of integrated pest management and sustainable food production systems has encouraged a greater interest in evaluating the effective utilization of cover crops in agricultural systems.

8.7.4 The Role of Cover Crops in Enhancing Biodiversity and Promoting Natural Pest Control Within Agricultural Systems

Cover crops play a crucial role in providing habitat and food sources for beneficial insects, pollinators, and soil organisms within agricultural systems. The presence of diverse cover crop mixtures offers a variety of plant species that serve as valuable resources for these beneficial organisms, contributing to enhanced biodiversity and natural pest control.

The abundance of flowering cover crops provides a rich source of nectar and pollen, attracting a diverse array of pollinators such as bees, butterflies, and other beneficial insects (Lu et al. 2000). These pollinators are essential for crop productivity and successful reproduction in many agricultural systems. Moreover, cover crops attract a range of predatory insects, including ladybugs, lacewings, and hoverflies, which prey on pests like aphids, caterpillars, and mites, helping to naturally regulate pest populations (Power 2010; Andrew et al. 2012; Deutsch et al. 2018; Sehgal et al. 2018). The diverse plant species within cover crop mixtures offer shelter, alternative prey, and nectar for these predatory insects, encouraging their establishment and population growth (Sehgal et al. 2018).

The integration of cover crops also supports soil biodiversity and the activity of beneficial soil organisms. The roots of cover crops release exudates and organic matter into the soil, providing food sources and habitat for beneficial soil microorganisms, including bacteria, fungi, and earthworms. These beneficial soil organisms contribute to nutrient cycling, organic matter decomposition, and soil structure improvement, enhancing overall soil health and fertility (Lu et al. 2000; Andrew et al. 2012).

The adoption of cover crops in agricultural systems can have significant benefits for enhancing biodiversity and promoting natural pest control. The provision of diverse plant resources, shelter, and food sources for beneficial insects, pollinators, and soil organisms can lead to a more sustainable and resilient agricultural ecosystem.

8.7.5 The Role of Cover Crops in Nutrient Cycling and Carbon Sequestration

Cover crops play a vital role in nutrient management and carbon sequestration within agricultural systems. These plants have the ability to take up nutrients from the soil, effectively reducing the risk of nutrient leaching and runoff (Lu et al. 2000; Outreach 2007; Hobbs et al. 2021). Their roots penetrate the soil, extracting nitrogen, phosphorus, and other essential nutrients that would otherwise be susceptible to being washed away. When the cover crop biomass is incorporated into the soil or left as surface residue, it undergoes decomposition, releasing the captured nutrients back into the soil, sustaining soil fertility and reducing the reliance on synthetic fertilizers (Lu et al. 2000; Andrew et al. 2012).

In addition to their nutrient management benefits, cover crops contribute to carbon sequestration, an important strategy for mitigating climate change. Through photosynthesis, cover crops capture atmospheric carbon dioxide and convert it into organic matter, sequestering carbon in their biomass. When the cover crop residues are incorporated into the soil, a portion of the captured carbon is added to the soil organic matter, enhancing its carbon content. Increased soil organic carbon not only improves soil health and fertility, but also plays a crucial role in climate change mitigation by removing carbon dioxide from the atmosphere and storing it within the soil (Lu et al. 2000; Outreach 2007; Andrew et al. 2012; Hobbs et al. 2021).

The need to develop increasingly integrated pest management and sustainable food production systems has encouraged a greater interest in thoroughly evaluating the effective utilization of cover crops in agricultural systems (Andrew et al. 2012). Cover crops provide numerous environmental benefits that can promote the long-term viability of farmlands, including weed control, habitat provision for beneficial organisms, and improvement of soil structure and water-holding capacity (Andrew et al. 2012). Some cover crops, such as legumes, can even fix nitrogen and contribute to the nitrogen requirements of subsequent crops.

Cover crops contribute indirectly to overall soil health by catching nutrients before they can leach out of the soil profile or, in the case of legumes, by adding nitrogen to the soil (Liu et al. 2024). Their roots can help unlock some nutrients, converting them to more available forms. Additionally, cover crops provide habitat or a food source for important soil organisms, break up compacted layers in the soil, and help dry out wet soils, reducing the risk of soil erosion (Outreach 2007).

8.7.6 The Role of Cover Crops in Improving Water Management in Agricultural Systems

Cover crops play a significant role in enhancing water-use efficiency within agricultural systems by improving soil structure, increasing soil porosity, and enhancing water-holding capacity (Andrew et al. 2012). One of the primary benefits of cover crops is their ability to improve water infiltration into the soil (Lu et al. 2000;

Outreach 2007; Gabriel et al. 2019). The dense and extensive root systems of cover crops create channels and pores in the soil, facilitating the movement of water into the deeper layers (Lu et al. 2000; Gabriel et al. 2019). This increased infiltration rate helps to mitigate surface runoff and prevent soil erosion by allowing water to enter the soil more efficiently (Lu et al. 2000).

Moreover, cover crops contribute to the improvement of soil structure, which is crucial for water management (Lu et al. 2000). The roots of cover crops penetrate and bind soil particles, promoting the aggregation of soil particles into stable aggregates (Lu et al. 2000; Outreach 2007). This aggregation enhances soil porosity, creating pathways for water movement and increasing the capacity of the soil to hold and store water (Lu et al. 2000; Gabriel et al. 2019). Improved soil structure allows water to infiltrate more easily, reducing the risk of runoff and promoting better water distribution throughout the soil profile (Lu et al. 2000; Gabriel et al. 2019).

The extensive root systems of cover crops also play a significant role in enhancing water-holding capacity (Lu et al. 2000) As cover crop roots penetrate the soil, they create channels and open spaces that increase the porosity and storage capacity of the soil (Gabriel et al. 2019). These larger pore spaces act as reservoirs for water, allowing the soil to retain more moisture and increasing the availability of water for plant uptake during dry periods (Lu et al. 2000; Gabriel et al. 2019).

The need to develop increasingly integrated pest management and sustainable food production systems has encouraged a greater interest in thoroughly evaluating the effective utilization of cover crops in agricultural systems (Andrew et al. 2012). Cover crops contribute indirectly to overall soil health by catching nutrients before they can leach out of the soil profile or, in the case of legumes, by adding nitrogen to the soil. Their roots can even help unlock some nutrients, converting them to more available forms (Outreach 2007). Cover crops provide habitat or a food source for some important soil organisms, break up compacted layers in the soil, and help dry out wet soils.

8.8 The Benefits of No-Tillage Farming: An Agroecological Approach

No-tillage farming, also known as zero-tillage or direct seeding, has gained significant attention in recent years as a sustainable agricultural practice that offers a multitude of benefits to both the environment and farming communities. This farming approach, which involves minimal soil disturbance and the elimination of conventional plowing or tilling operations, is considered a key agroecological practice due to its positive impacts on soil health, water conservation, and biodiversity (Wilson and Lovell 2016).

One of the primary advantages of no-tillage farming is its ability to improve soil management and quality. By maintaining a permanent soil cover, no-tillage systems foster the development of a diverse and thriving soil ecosystem, which can enhance nutrient cycling, water infiltration, and the overall physical and chemical properties of the soil (Blevins and Frye 1993; Tamburini et al. 2016). This, in turn, can lead to increased crop yields and reduced input costs, as the soil becomes more resilient and self-sustaining.

Additionally, the elimination of intensive tillage operations helps to mitigate soil erosion, a major concern in many agricultural landscapes (Tamburini et al. 2016). By maintaining a protective layer of crop residues on the soil surface, no-tillage farming reduces the risk of soil loss due to wind and water, preserving the long-term productivity of the land.

Furthermore, no-tillage farming has been shown to improve water conservation and quality. The absence of soil disturbance and the presence of a permanent vegetation cover can enhance water infiltration, reduce surface runoff, and limit the transport of agrichemicals and sediments into nearby water bodies.

This practice can also contribute to the enhancement of on-farm biodiversity, as it provides a more stable and diverse habitat for a range of organisms, including beneficial insects, soil microorganisms, and even small vertebrates.

The adoption of no-tillage farming has also been linked to various socioeconomic benefits, such as reduced labor and fuel costs, improved farm profitability, and increased food security for local communities (Tamburini et al. 2016; Arriaga et al. 2017).

As the global agricultural sector faces the pressing challenges of climate change, resource depletion, and environmental degradation, the adoption of no-tillage farming represents a promising agroecological approach that can help to address these challenges and promote more sustainable and resilient food production systems (Blevins and Frye 1993; Mijangos et al. 2006; Tamburini et al. 2016).

8.8.1 The Role of No-Tillage in Soil Conservation

No-tillage, a conservation practice that leaves crop residues or cover crops on the soil surface, plays a crucial role in preventing soil erosion within agricultural systems. This approach helps protect the soil by improving water infiltration, preserving soil structure, and preventing the loss of topsoil (Blevins and Frye 1993).

One of the primary benefits of no-tillage is the preservation of crop residues or cover crops on the soil surface (Marioti et al. 2013). Instead of tilling the soil and burying crop residues, these residues are left undisturbed on the surface. The crop residues act as a protective layer, shielding the soil from the impact of raindrops and reducing the erosive power of rainfall (Blevins and Frye 1993). This protective layer of residue helps to dissipate the energy of raindrops, preventing soil particles from being detached and transported by runoff.

The presence of crop residues on the soil surface also contributes to improved water infiltration. The residue layer acts as a physical barrier, reducing the impact of raindrops and preventing the formation of surface crusts, which can impede water infiltration. By enhancing water infiltration, no-tillage helps to promote a more efficient use of rainfall, allowing water to penetrate into the soil and replenish soil moisture (Blevins and Frye 1993).

In addition to improving water infiltration, no-tillage preserves soil structure by minimizing soil disturbance. Traditional tillage practices can disrupt soil aggregates and destroy soil structure, making the soil more susceptible to erosion. In contrast, no-tillage avoids the disturbance of the soil, allowing the natural structure and integrity of the soil to be maintained. This preserved soil structure enhances the stability of soil aggregates, promoting good porosity and water-holding capacity (Blevins and Frye 1993; Tamburini et al. 2016). By preventing soil erosion, no-tillage also helps to preserve the topsoil, a critical component of soil fertility and productivity.

8.8.2 The Role of No-Tillage in Enhancing Soil Health and Fertility

No-tillage, a widely adopted conservation practice, has emerged as a crucial strategy in promoting the accumulation of organic matter in the soil, which is a cornerstone of soil health and fertility. By leaving crop residues on the soil surface, no-tillage practices provide a continuous source of organic material, gradually contributing to the buildup of soil organic carbon. Soil organic carbon is a critical component of soil health, as it plays a significant role in improving soil structure, nutrient cycling, and water-holding capacity (Tamburini et al. 2016).

The retention of crop residues on the soil surface through no-tillage practices enhances nutrient cycling within the soil. As the crop residues decompose, they release essential nutrients back into the soil, making them available for subsequent crop uptake, thereby contributing to the long-term fertility and productivity of the soil (Blevins and Frye 1993). Furthermore, no-tillage supports beneficial soil microbial activity, including earthworms and other organisms, which further contribute to nutrient availability and overall soil health (Swami 2020).

Numerous studies have demonstrated the positive impact of no-tillage on soil health and fertility. The incorporation of organic matter into the soil through no-tillage practices has been shown to improve soil structure, enhance water infiltration, and increase the soil's water-holding capacity. Additionally, the reduced disturbance of the soil in no-tillage systems has been found to promote the growth of beneficial soil organisms, such as earthworms, which play a vital role in improving soil structure and nutrient cycling (Blevins and Frye 1993; Mijangos et al. 2006; Tamburini et al. 2016; Nayak et al. 2020).

8.8.3 Water Conservation and Efficiency: The Role of No-Tillage Practices

No-tillage practices have a significant impact on reducing evaporation and water loss from the soil, thereby improving water-use efficiency within agricultural systems (Blevins and Frye 1993; Tamburini et al. 2016). By leaving crop residues or cover crops on the soil surface, no-tillage practices create a protective cover that acts as a barrier against evaporation, reducing the rate of evaporation and preventing excessive water loss. In dry and arid regions where water availability is limited, no-tillage plays a crucial role in improving water-use efficiency by minimizing evaporation and promoting the retention of soil moisture, ensuring a more efficient use of limited water resources (Marioti et al. 2013; Tamburini et al. 2016). This protective cover helps to create a microclimate that reduces soil surface temperature and wind speed, further contributing to the conservation of soil moisture (Tamburini et al. 2016). The retention of soil moisture facilitated by no-tillage practices offers several benefits in agricultural systems, including maintaining adequate soil moisture levels for plant growth, particularly during dry spells or drought conditions, and increasing water-use efficiency as a larger proportion of the water is retained within the root zone and available for plant uptake (Marioti et al. 2013; Kumawat et al. 2020). Moreover, effective rainwater conservation and efficient water use can be accomplished through techniques such as conservation tillage, mulch cultivation, limited irrigation, and the introduction of drought-tolerant varieties and crop rotation systems (Tamburini et al. 2016; Kumawat et al. 2020).

8.8.4 Carbon Sequestration and Climate Change Mitigation Through No-Tillage Practices

The role of agricultural practices in mitigating climate change has gained significant attention, with no-tillage emerging as a promising approach to enhancing carbon sequestration in soils. No-tillage, by minimizing soil disturbance and preserving crop residues, facilitates the accumulation of organic carbon, thereby contributing to the reduction of greenhouse gas emissions and promoting long-term soil carbon storage (Morgan et al. 2010; Hobbs et al. 2021).

One of the key mechanisms through which no-tillage promotes carbon sequestration is by reducing soil disturbance. Traditional tillage practices involve mechanical soil disturbance, which accelerates the decomposition of organic matter and releases carbon dioxide (CO_2) into the atmosphere (Morgan et al. 2010). In contrast, no-tillage practices minimize soil disturbance by leaving crop residues on the soil surface and avoiding the inversion of soil layers. This helps to preserve the integrity of soil aggregates and reduce carbon loss through increased decomposition (Bolinder et al. 2020).

The preservation of crop residues in no-tillage systems also contributes significantly to increased carbon storage in the soil. Crop residues, such as plant residues and cover crops, contain substantial amounts of carbon. By leaving these residues on the soil surface, no-tillage practices provide a continuous source of organic material that undergoes decomposition, leading to the gradual accumulation of soil organic carbon. The retention of crop residues also enhances soil organic matter content, improving soil structure, nutrient availability, and water-holding capacity, which further contribute to long-term carbon sequestration in the soil (Chenu et al. 2019; Bolinder et al. 2020).

The accumulation of organic carbon in the soil through no-tillage practices has significant implications for climate change mitigation. Carbon sequestration in the soil helps to offset greenhouse gas emissions, making no-tillage a valuable strategy in the overall effort to mitigate the impacts of climate change (Chenu et al. 2019).

8.8.5 Biodiversity and Wildlife Habitat: The Benefits of No-Tillage Systems

No-tillage systems offer numerous advantages for promoting biodiversity and supporting beneficial insects, earthworms, and other soil organisms within agricultural landscapes. The presence of crop residues on the soil surface in these systems creates a favorable habitat and refuge for a wide range of beneficial organisms, enhancing ecosystem balance and natural pest control (Blevins and Frye 1993; Tamburini et al. 2016; Cozim-Melges et al. 2024).

The crop residues in no-tillage systems serve as protective cover, providing a suitable environment for the establishment and activity of beneficial insects, such as predators and parasitoids (Cozim-Melges et al. 2024). This can lead to an increase in the abundance and diversity of these insects, which play a crucial role in natural pest control by preying on agricultural pests (Tamburini et al. 2016). Earthworms, as ecosystem engineers, also thrive in the presence of crop residues, which provide them with organic matter and a food source, stimulating their activity and population growth (Tamburini et al. 2016). Earthworms are essential for soil health and nutrient cycling processes, such as organic matter decomposition and nutrient availability, thus contributing to the overall sustainability and productivity of agricultural systems (Blevins and Frye 1993; Tamburini et al. 2016; Wilson and Lovell 2016).

In addition to supporting beneficial insects and earthworms, no-tillage systems provide refuge and habitat for a variety of other wildlife species. The reduction in chemical inputs associated with no-tillage practices can have positive effects on wildlife populations by minimizing their exposure to pesticides (Blevins and Frye 1993; Tamburini et al. 2016; Wilson and Lovell 2016).

While adapting no-till techniques for organic farming can be challenging, particularly in water-limited environments, the benefits of no-tillage systems for promoting biodiversity and supporting beneficial organisms make it a valuable approach to sustainable agriculture (Blevins and Frye 1993; Tamburini et al. 2016; Wilson and Lovell 2016).

8.8.6 The Environmental and Economic Benefits of No-Tillage Systems

No-tillage systems offer a suite of advantages over conventional tillage practices, including substantial reductions in fuel consumption and greenhouse gas emissions, as well as significant cost savings in terms of labor and machinery requirements.

One of the primary benefits of no-tillage systems is the reduction in fuel usage and the associated decrease in greenhouse gas emissions. Compared to traditional tillage methods that involve multiple passes with heavy machinery, no-tillage systems require fewer field operations, resulting in lower fuel consumption (Blevins and Frye 1993; Tamburini et al. 2016; Šarauskis et al. 2018). This directly translates into diminished emissions of carbon dioxide (CO_2), nitrous oxide (N_2O), and methane (CH_4)—greenhouse gases that contribute to climate change. By minimizing soil disturbance and the need for frequent machinery use, no-tillage systems help mitigate the environmental impact of agricultural practices.

Furthermore, no-tillage systems offer substantial economic advantages in the form of reduced labor and machinery costs. Traditional tillage requires numerous passes with tillage equipment, which necessitates additional investment in both labor and machinery. In contrast, no-tillage systems eliminate or significantly reduce the need for primary tillage, seedbed preparation, and other soil cultivation activities, thereby lowering labor requirements (Šarauskis et al. 2018). The reduced labor and machinery costs associated with no-tillage make it a more economically viable option for farmers, particularly in terms of long-term sustainability and profitability (Tamburini et al. 2016). A study by Šarauskis et al. (2018) found that the use of sustainable tillage practices and soil preparation without a plow can reduce costs by 25% to 41% compared to conventional tillage.

The environmental and economic advantages of no-tillage systems extend beyond just fuel and cost savings. By requiring fewer passes with machinery, no-tillage systems also contribute to soil conservation by minimizing soil compaction and preserving soil structure. Maintaining soil structure is crucial for sustaining soil fertility, water infiltration capacity, and nutrient availability—all of which are essential for long-term agricultural productivity and sustainability.

8.9 Integrating Fertilizer Application and Fertigation into Agroecological Practices in Sub-Saharan Africa

Agroecological practices in sub-Saharan Africa have the potential to enhance crop productivity while minimizing environmental impacts, and the strategic use of fertilizer application and fertigation (fertilizer application through irrigation systems) can play a key role in achieving this balance (Nyamangara et al. 2020). Fertilizer subsidies have been an area of debate, with some experts arguing they are essential for long-term food security, while others raise concerns about the market distortions they can introduce. Regardless, the application of synthetic fertilizers in the region remains low, often due to limited access and resources for smallholder farmers. Furthermore, the negative environmental impacts observed in these smallholder systems are typically attributed to nutrient mining and land degradation, rather than excessive fertilizer use.

Women farmers, who do the majority of food production in sub-Saharan Africa, face additional challenges in accessing the resources needed to improve soil fertility, including cash, credit, and both organic and inorganic fertilizers. Addressing these gendered disparities in access to agricultural inputs and information will be crucial for realizing the potential of fertilizer application and fertigation within agroecological approaches.

Fertigation system combining fertilization and irrigation to deliver nutrients to plants through irrigation water

Targeted fertilizer subsidy programs may offer a pathway to increase fertilizer use and improve soil fertility, while minimizing the market distortions associated with universal subsidy schemes. Integrated soil fertility management approaches, such as the Soil Health Project by the Alliance for a Green Revolution in Africa, have also demonstrated the potential to boost land productivity and agricultural incomes sustainably (Martey et al. 2019). By carefully designing and implementing fertilizer application and fertigation strategies that are tailored to the local agroecological context and address the specific needs of smallholder farmers, sub-Saharan Africa can harness these practices to enhance crop productivity and environmental stewardship (Druilhe and Barreiro-Hurlé 2012; Martey et al. 2019; Nyamangara et al. 2020).

8.10 Balanced Nutrient Management in Agroecology

Agroecology emphasizes the use of organic and mineral fertilizers in a balanced and judicious manner to promote sustainable and environmentally friendly agricultural practices. This text will elaborate on the importance of nutrient management in agroecology, including the use of soil nutrient analysis, crop nutrient requirements, and nutrient cycling principles. It will also discuss precise and targeted fertilizer application techniques, such as fertigation, that optimize nutrient use efficiency and minimize environmental impacts.

In agroecology, the application of fertilizers is approached with careful consideration of soil nutrient analysis and crop nutrient requirements. Soil nutrient analysis helps determine the existing nutrient levels in the soil, enabling farmers to make informed decisions regarding fertilizer application (Ausari et al. 2020). By understanding the nutrient composition of the soil, farmers can tailor their fertilizer use to address specific nutrient deficiencies and maintain optimal nutrient levels for crop growth (Ausari et al. 2020). Crop nutrient requirements play a crucial role in agroecological fertilizer management, as different crops have varying nutrient demands at different growth stages. By understanding the specific nutrient needs of each crop, farmers can apply fertilizers in a targeted manner, optimizing nutrient uptake and minimizing excess nutrient application. Furthermore, agroecology emphasizes the principles of nutrient cycling, which involves recycling nutrients within the agricultural system. Nutrient cycling involves the efficient use of organic matter, such as crop residues and animal manure, to provide a source of nutrients for future crops (Kumar et al. 2018a, b, c; Ausari et al. 2020; Havlin and Heiniger 2020).

8.11 The Role of Organic Fertilizers in Agroecology

Agroecological practices place a strong emphasis on the use of organic fertilizers as a key component of sustainable and environmentally friendly agriculture. Organic fertilizers, such as compost, manure, and crop residues, offer numerous benefits for soil fertility, nutrient availability, soil structure, microbial activity, and long-term soil health (Volkohon et al. 2016; Patel and Champaneri 2020).

One of the primary advantages of organic fertilizers is their slow-release nature, which ensures a more balanced nutrient supply to plants over time. This characteristic reduces the risk of nutrient imbalances and minimizes nutrient losses to the environment, aligning with the principles of agroecology. Organic matter in fertilizers also plays a crucial role in improving soil structure, enhancing soil aggregation and stability, and increasing soil porosity and water infiltration (Swami 2020).

Organic fertilizers serve as a food source for beneficial soil microorganisms, including bacteria and fungi, which contribute to nutrient cycling and overall soil health (Verma et al. 2017; Dotaniya et al. 2020; Nayak et al. 2020; Swami 2020). The presence of organic matter stimulates microbial growth and activity, leading to increased nutrient mineralization and availability for plant uptake (Verma et al. 2017; Dotaniya et al. 2020; Nayak et al. 2020; Swami 2020).

Furthermore, the continuous application of organic fertilizers fosters long-term soil health, as they help maintain and enhance soil organic matter levels, which are crucial for the sustainability of agroecological systems.

8.11.1 Nutrient Cycling and Sustainable Soil Management in Agroecology

Agroecology, a holistic approach to agricultural management, places a strong emphasis on the importance of nutrient cycling within farming systems as a means to enhance soil fertility and minimize nutrient losses. This paper explores the significance of nutrient cycling in agroecology and how practices such as fertilization and fertigation contribute to sustainable nutrient management.

Nutrient cycling is a fundamental concept in agroecology, involving the efficient utilization and recycling of nutrients within farming systems. Rather than relying solely on external inputs, agroecological practices aim to optimize nutrient cycling by applying fertilizers and amendments in a manner that promotes their utilization and minimizes losses. By adopting strategies that enhance nutrient cycling, farmers can improve soil fertility and reduce the risk of nutrient losses through leaching or runoff (Fontaine et al. 2024). Nutrient cycling involves the incorporation of organic matter, such as crop residues, cover crops, and organic fertilizers, into the soil. These organic materials serve as nutrient sources, which are gradually released and made available to plants over time through decomposition and microbial activity (Edwards et al. 1993).

Fertigation, a combination of fertilization and irrigation, is an effective technique employed in agroecology to deliver nutrients directly to the root zone of plants. By applying nutrients through irrigation water, fertigation allows for precise and targeted nutrient application, minimizing the risk of nutrient leaching or runoff as the nutrients are efficiently taken up by the crops in the root zone (Raimi et al. 2017; Barman et al. 2019). The integration of fertigation in agroecological systems provides several benefits, including the ability to deliver nutrients in a controlled manner, matching the specific nutrient requirements of the crops at different growth stages. This precise nutrient management minimizes the excess application of nutrients, which can lead to environmental degradation or economic waste.

The importance of nutrient cycling in agroecology is further supported by research on the potential of bio-fertilizers and integrated nutrient management. Increasing crop productivity per unit of land area to meet future food and fiber demand requires replenishing soil fertility through efficient nutrient management practices (Havlin and Heiniger 2020). Bio-fertilizers, composed of beneficial microorganisms, have shown promise in enhancing nutrient availability, plant growth, and crop yields in a sustainable manner (Raimi et al. 2017; Barman et al. 2019).

8.11.2 *Optimizing Nutrient Management: The Benefits of Fertigation and Precision Agriculture*

Fertigation is a valuable technique in agroecology that enables farmers to apply fertilizers with precision and control. This technique offers numerous benefits, including precise nutrient dosing, resource efficiency, and the role of precision agriculture technologies in optimizing nutrient management.

Fertigation provides farmers the ability to apply fertilizers in a precise and controlled manner, allowing for optimal nutrient dosages based on the specific needs and growth stages of crops (Senthilkumar et al. 2017). By integrating fertilization with irrigation, fertigation offers a convenient and efficient approach to deliver nutrients directly to the root zone of plants (Senthilkumar et al. 2017). Fertigation has been shown to result in more than 40% savings in fertilizer use while also increasing yields and improving product quality (Nirala et al. 2018). The precise application of fertilizers through fertigation allows farmers to avoid over-application and reduce nutrient losses. With traditional fertilization methods, there is a higher risk of nutrient runoff or leaching, which can lead to water pollution and waste of valuable resources. Fertigation minimizes these risks by ensuring that nutrients are applied in the right quantities and at the right time, closely matching the crop's requirements (Senthilkumar et al. 2017; Nirala et al. 2018).

Furthermore, precision agriculture technologies play a crucial role in enhancing nutrient management and improving resource-use efficiency. Soil sensors and real-time monitoring systems enable farmers to gather accurate data on soil nutrient levels, moisture content, and other relevant parameters. By using this information, farmers can make informed decisions about the timing and dosage of fertigation, ensuring that nutrients are applied when and where they are most needed. This data-driven approach maximizes nutrient uptake by crops and minimizes wastage. Precision agriculture technologies also facilitate site-specific variable rate application of fertilizers, further optimizing resource use and minimizing environmental impacts (Loon et al., 2018).

8.11.3 *The Environmental Advantages of Fertigation in Agroecological Systems*

Agroecology, as an ecological approach to agriculture, emphasizes the minimization of negative environmental impacts and the promotion of sustainable practices. One such practice that aligns with the goals of agroecology is fertigation, which, when properly managed, can effectively reduce the risk of nutrient runoff and water pollution compared to conventional surface-applied fertilizers. Fertigation achieves this by delivering nutrients directly to the root zone of plants, minimizing the potential for nutrient loss through runoff.

Conventional fertilizer application often leads to the issue of nutrient runoff, where excess nutrients are carried away by rainfall or irrigation water and enter water bodies, leading to water pollution, eutrophication, and detrimental impacts on aquatic ecosystems (Nirala et al. 2018) Fertigation offers a solution to this problem by enabling the targeted and precise application of nutrients, ensuring that they are applied only where they are needed by the crops. By integrating fertilization with irrigation, fertigation allows farmers to apply nutrients in smaller, more frequent doses, reducing the risk of excessive nutrient application and subsequent runoff (Barman et al. 2019; Srivastav 2020).

This targeted approach helps to minimize fertilizer contact with non-target areas, such as water bodies or adjacent ecosystems, thereby reducing the environmental risks associated with nutrient pollution (Carrasco and Urrestarazu 2010; Barman et al. 2019). Moreover, fertigation can enhance nutrient use efficiency, as it enables the synchronization of nutrient supply with the crop's demand. By delivering nutrients directly to the root zone in a readily available form, fertigation maximizes nutrient uptake by the plants, reducing the amount of unused nutrients that could potentially be lost to the environment (Carrasco and Urrestarazu 2010; Senthilkumar et al. 2017; Nirala et al. 2018). The potential benefits of fertigation in reducing environmental impact have been well-documented in the literature. One study found that fertigation can result in a 25–50% reduction in fertilizer use compared to conventional application methods, while also improving water-use efficiency (Nirala et al. 2018). Another review highlighted the ability of fertigation to deliver nutrients directly to the root zone, minimizing nutrient losses and promoting higher productivity (Senthilkumar et al. 2017).

8.12 Climate Resilience Through Agroecological Practices: The Role of Efficient Fertilizer Application Techniques

Agroecological practices, including efficient fertilizer application techniques like fertigation, play a significant role in promoting climate resilience in agricultural systems (Altieri and Nicholls 2013). These practices contribute to enhanced crop productivity, improved soil health, and the mitigation of climate change impacts (Ghasemzadeh 2012; Altieri and Nicholls 2013; Maia et al. 2018).

Efficient fertilizer application techniques, such as fertigation, deliver nutrients directly to the root zone of plants, greatly enhancing crop productivity. By precisely meeting the nutrient demands of crops, fertigation optimizes nutrient uptake and utilization, leading to improved plant growth, yield, and overall crop performance (Ghasemzadeh 2012). This enhanced productivity is essential for building climate resilience in agricultural systems, as it allows farmers to sustainably meet food demands while adapting to changing climatic conditions.

Agroecological practices, including appropriate nutrient application, contribute to crop tolerance to climate stressors. Well-managed and balanced nutrient supply helps plants develop stronger root systems, efficient photosynthesis, and improved stress tolerance mechanisms. This enables crops to better withstand abiotic stressors such as drought, heat, and salinity, which are becoming increasingly common under climate change scenarios. By supporting crop resilience, efficient nutrient application techniques like fertigation can help mitigate the negative impacts of climate change on agricultural productivity (Ghasemzadeh 2012).

Furthermore, agroecological practices that prioritize soil health, including judicious fertilizer application, have the potential to contribute to carbon sequestration in soils. Carbon sequestration refers to the process of capturing and storing atmospheric carbon dioxide in the soil, thereby mitigating its release into the atmosphere and mitigating climate change. Fertigation, when combined with other agroecological practices that enhance soil organic matter and biodiversity, can play a role in increasing soil carbon storage and reducing greenhouse gas emissions from agricultural systems (Altieri and Nicholls 2013; Altieri et al. 2015; Havlin and Heiniger 2020; Fontaine et al. 2024).

8.13 The Role of Permaculture in Sustainable Development in Sub-Saharan Africa

Permaculture, a holistic design approach that mimics natural ecosystems, holds immense potential for driving sustainable development in sub-Saharan Africa. The region's heavy reliance on agriculture makes it particularly well-suited for the adoption of permaculture practices, which can enhance food production, ecological restoration, and community resilience (Abdulai and Hazell 1996; Tenywa and Bekunda 2009; Leakey 2020).

Sub-Saharan Africa faces significant challenges in managing its soil and water resources, with issues such as destructive extraction, over-exploitation, and inadequate conservation leading to land degradation and reduced agricultural productivity (Tenywa and Bekunda 2009; Leakey 2020). Permaculture offers a viable solution by promoting the rehabilitation of degraded farmland through the integration of diverse, nutrient-dense crops and the restoration of agroecological functions (Leakey 2020). Furthermore, permaculture principles emphasize the importance of water harvesting and utilization, which can help address the region's water scarcity challenges and increase the resilience of local communities to the impacts of climate change.

Success stories of local soil and water conservation activities practiced by farmers in sub-Saharan Africa demonstrate the potential of permaculture, but the challenge lies in scaling these successes to wider implementation. Ecohydrology, a key component of permaculture, can play a crucial role in this process by transferring knowledge and practical applications to local communities, thereby enhancing their capacity to manage natural resources sustainably.

In addition to its environmental benefits, permaculture also has the potential to drive social and economic development in sub-Saharan Africa. By diversifying and rehabilitating degraded farmland, permaculture can improve food production, create new local business opportunities, and enhance social well-being, ultimately promoting livelihoods and social justice (Abdulai and Hazell 1996; Tenywa and Bekunda 2009; Leakey 2020; Leakey 2020).

Women, who make up a significant portion of the agricultural labor force in the region, can play a crucial role in the adoption and implementation of permaculture practices. Empowering women to participate in soil and water conservation efforts can lead to more sustainable and equitable outcomes, contributing to the overall resilience and development of communities in sub-Saharan Africa.

A permaculture farm integrating agricultural practices with natural ecosystems

The region's challenges in managing soil and water resources, such as destructive extraction and over-exploitation, have led to significant land degradation and reduced agricultural productivity. Permaculture, with its emphasis on ecosystem restoration and sustainable resource management, offers a promising solution to these issues. By integrating diverse, nutrient-dense crops and restoring agroecological functions, permaculture can enhance food production and increase the resilience of local communities to the impacts of climate change.

8.13.1 Agroecological Design: Integrating Permaculture Principles for Sustainable Agriculture in Sub-Saharan Africa

Permaculture design principles advocate for working in harmony with nature, and in sub-Saharan Africa, this approach aims to create resilient and productive agro-ecosystems that are tailored to local climatic conditions, soil types, and natural resources (Francis et al. 2003). The core concept of permaculture is to design and integrate diverse and interrelated elements, such as trees, shrubs, crops, livestock, water management systems, and infrastructure, to create self-sustaining and regenerative systems that enhance overall resilience and productivity (Leakey 2020). One key principle of permaculture design is observation, where practitioners carefully study the natural patterns and processes occurring in the local environment, such as rainfall patterns, sunlight exposure, and wind direction, to inform the design of the agroecosystem and optimize the placement of different elements. Another fundamental principle is integration, where permaculture design seeks to create functional relationships between different components, maximizing the synergies and positive interactions between them to result in a more resilient and productive system. Permaculture also emphasizes the concept of diversity, incorporating a wide variety of plants, animals, and microorganisms to mimic the diversity found in natural ecosystems, which helps create a more resilient system as different species perform complementary functions (Edwards et al. 1993; Chirinda et al. 2017; Jayne et al. 2018; Leakey 2020).

The maintenance of biological diversity and nutrient cycling mechanisms are global principles that are common to all agroecosystems and therefore essential in the design of sustainable agricultural systems. However, regional or site-specific factors, such as climate, soils, and socioeconomic preferences and conditions, can differ greatly among agroecosystems and may assume major importance in some. Interdisciplinarity, participation of farmers, and a whole-farm level approach are fundamental to research and development on these global commonalities and regional factors. Agroecology has evolved from an ecology-based discipline with five principles to a broader, multidimensional concept that incorporates social, political, and economic dimensions (Barrios et al. 2020).

8.13.2 Food Security and Sustainable Agriculture: The Role of Permaculture

Permaculture places a strong emphasis on food security by promoting diversified and sustainable agricultural practices (Pinstrup-Andersen and Herforth 2008; Siegner et al. 2018). By integrating a variety of food crops, fruit trees, and livestock, permaculture systems aim to provide a diverse and nutritious food supply (Frelat et al. 2016). This approach recognizes the importance of agrobiodiversity in enhancing resilience and reducing the vulnerability of food production systems to pests, diseases, and environmental shocks (Frelat et al. 2016).

One key aspect of permaculture is the promotion of agrobiodiversity. Unlike conventional monoculture systems, permaculture encourages the cultivation of multiple crop varieties with different growth habits, nutrient requirements, and resistance to pests and diseases. This diversity not only helps to spread risks but also enhances ecosystem resilience, as a diverse range of crops attracts beneficial insects, pollinators, and natural predators, contributing to natural pest control and reducing the dependence on synthetic pesticides (Allen et al. 2014; Ansari et al. 2023). Furthermore, permaculture practices focus on improving soil fertility through natural means. Techniques such as composting, mulching, and the use of cover crops help increase organic matter content in the soil, enhance nutrient cycling, and improve soil structure. These practices support the development of healthy soil ecosystems and nutrient-rich soils, which are vital for the production of high-quality and nutrient-dense crops. Water conservation is another important aspect of permaculture. Through efficient irrigation methods, such as drip irrigation, swales, and rainwater harvesting, permaculture systems minimize water loss and maximize water-use efficiency. By capturing and storing water on-site, permaculture systems can reduce the dependence on external water sources and mitigate the impacts of drought and water scarcity (Chirinda et al. 2017; Forister et al. 2019; Sethuraman et al. 2021).

Drip irrigation lines laid on a lettuce crop to efficiently deliver water directly to the root zone, reducing water wastage and promoting optimal plant growth

Permaculture also seeks to minimize the use of synthetic inputs, such as chemical fertilizers and pesticides. Instead, it promotes the integration of biological and ecological processes to enhance productivity and resilience (Delgado et al. 2019;

Forister et al. 2019). Recent research has shown that adopting practices that mimic natural ecosystems, such as increasing plant diversity, enhancing soil organic matter, and reducing tillage, can improve nutrient cycling, reduce greenhouse gas emissions, and increase carbon sequestration in agricultural systems (Forister et al. 2019).

8.13.3 Sustainable Water Management Strategies in Sub-Saharan Africa: The Role of Permaculture

Water scarcity is a pressing challenge in many parts of sub-Saharan Africa, and permaculture design principles offer valuable strategies for managing water resources sustainably (Larsen et al. 2016). Permaculture aims to mitigate the impacts of water scarcity and ensure the availability of water for agricultural activities, particularly during dry periods, through the incorporation of various water management techniques (Tenywa and Bekunda 2009; Stroosnijder et al. 2012).

One key approach in permaculture is rainwater harvesting, which involves the collection and storage of rainwater for later use. Permaculture systems incorporate infrastructure such as roof catchment systems, gutters, and downspouts to direct rainwater into storage tanks or ponds. This stored water can then be used for irrigation, livestock watering, and other agricultural needs, reducing reliance on scarce groundwater sources or surface water bodies (Tenywa and Bekunda 2009; Stroosnijder et al. 2012; Larsen et al. 2016; Shadeed et al. 2020).

Swales are another water management technique commonly used in permaculture. Swales are contour ditches or trenches that follow the natural slope of the land, acting as passive water-harvesting structures that capture and direct rainwater to infiltrate the soil, thereby increasing water availability in the root zone of plants (Stroosnijder et al. 2012). By slowing down the flow of water, swales also help prevent erosion and promote the recharging of groundwater (Stroosnijder et al. 2012).

Contouring, which involves shaping the land along contour lines to create level or gently sloping areas, is closely related to swales and helps retain water on the land, allowing it to infiltrate the soil and benefit plant growth (Hudson 1987; Stroosnijder et al. 2012; Shadeed et al. 2020). Contouring also reduces water runoff, which can carry away valuable topsoil and nutrients (Stroosnijder et al. 2012).

The use of ponds or tanks for water storage is another important aspect of permaculture water management. These structures allow for the collection and storage of larger volumes of water, providing a reliable source of irrigation water during dry periods (Hudson 1987; Stroosnijder et al. 2012). Ponds can also serve as habitat for aquatic organisms, supporting biodiversity within the agricultural landscape (Hudson 1987; Rao 2005; Stroosnijder et al. 2012; Gebreyess and Amare 2019; Shadeed et al. 2020).

8.13.4 Soil Conservation and Regeneration in Permaculture Systems: Strategies for Sustainable Agriculture in Sub-Saharan Africa

Permaculture places significant emphasis on practices that prioritize soil health and fertility, recognizing the essential role of soil in the long-term sustainability and productivity of agricultural systems in sub-Saharan Africa. One of the key practices promoted in permaculture for improving soil health is agroforestry, which involves the intentional integration of trees with agricultural crops or livestock. sentence "The presence of trees in agroforestry…" for clarity. Replace with: The presence of trees in agroforestry systems provides various benefits to the soil, including stabilizing it, reducing erosion and nutrient leaching, and enriching it with nutrients through leaf litter and organic matter (Sanchez et al. 1997; Nair 2014). Another essential technique in permaculture for supporting soil health is cover cropping, which involves planting cover crops between main crops or during fallow periods to protect the soil surface from erosion, fix atmospheric nitrogen, increase organic matter content, and enhance soil structure (Sanchez et al. 1997; Tenywa and Bekunda 2009; Nair 2014; Koskey et al. 2021). Mulching, where organic materials are spread on the soil surface, is also commonly employed in permaculture systems to protect the soil, conserve moisture, and promote soil fertility by providing a source of decomposing organic matter. Composting, through which organic materials are transformed into nutrient-rich soil amendments, is another important technique advocated by permaculture for improving soil fertility (Sanchez et al. 1997; Nair 2014).

8.13.5 Energy Efficiency and Renewable Energy in Permaculture Systems

Permaculture places a strong emphasis on energy efficiency and promoting sustainable practices that minimize energy inputs and maximize the utilization of renewable energy sources. By incorporating various design principles and technologies, permaculture systems in sub-Saharan Africa aim to reduce dependence on fossil fuels and promote the efficient use of energy.

Passive solar design principles are often employed in permaculture to optimize the use of solar energy for heating, cooling, and lighting in buildings. Passive solar design utilizes building orientation, insulation, thermal mass, and natural ventilation to maximize natural heating and cooling, reducing the need for mechanical heating or cooling systems and saving energy. Solar energy is abundantly available in sub-Saharan Africa, with Senegal, for example, receiving an average of 5.8 kWh/m of solar energy per day (Cisse et al. 2023). The energy-intensive nature of buildings in Africa, particularly due to high use of ventilation and air conditioning for thermal comfort, makes passive solar design a particularly valuable approach.

Utilizing renewable energy sources is another key aspect of permaculture's energy efficiency focus. Solar panels can be used to harness solar energy and convert it into electricity for various purposes, such as lighting, powering appliances, or pumping water. This reduces reliance on non-renewable energy sources and minimizes greenhouse gas emissions associated with electricity generation. Solar photovoltaic (PV) technology has had a significant impact on rural electrification in developing countries, though the current status of solar home systems in sub-Saharan Africa is not well-documented. Opportunities exist for further cost reductions through local manufacturing of solar components.

Biogas digesters are another technology promoted in permaculture to utilize renewable energy. Biogas digesters convert organic waste, such as animal manure or crop residues, into biogas, a renewable energy source that can be used for cooking, heating, or generating electricity. Biogas digesters not only provide clean energy but also contribute to waste management and the reduction of methane emissions from decomposing organic matter.

Reducing reliance on fossil fuels for farming operations is also emphasized in permaculture. This can be achieved through various practices, such as utilizing human or animal power for fieldwork, employing efficient irrigation systems that minimize water and energy use, and adopting agroecological practices that reduce the need for synthetic inputs and machinery. Integrating solar energy into drip irrigation systems has been shown to enhance food security in sub-Saharan Africa by conserving water, enhancing reliability of power, and conserving land and space (Hernandez et al. 2015; Cisse et al. 2023).

A biogas system for energy generation from kitchen waste

Overall, permaculture's emphasis on energy efficiency and renewable energy sources has significant potential to address the energy challenges facing sub-Saharan Africa, reducing greenhouse gas emissions, enhancing food security, and promoting sustainable development.

8.13.6 Community Engagement and Knowledge Sharing: The Transformative Potential of Permaculture in Sub-Saharan Africa

Permaculture, a holistic approach to sustainable agriculture, has gained significant attention for its emphasis on community engagement, participation, and knowledge sharing. In the context of sub-Saharan Africa, this integrated system offers a promising solution to address various challenges faced by rural communities, including food security, water scarcity, soil degradation, and climate resilience.

At the heart of permaculture lies the principle of fostering community-driven initiatives and collaborative learning networks. By actively involving community members in the design, implementation, and management of permaculture systems, it not only enhances collective problem-solving capacity but also strengthens the sense of ownership and empowerment within the community. This approach aligns with the growing recognition that successful interventions in rural development must be grounded in local knowledge, resources, and cultural practices to achieve long-term sustainability (Leakey 2020).

One of the key features of permaculture is the integration of diverse and interconnected elements within agricultural systems. By incorporating a variety of crops, trees, livestock, and beneficial organisms, permaculture systems mimic natural ecosystems and enhance biodiversity. This diversity not only increases overall system resilience but also provides multiple sources of food, income, and ecological services (Tenywa and Bekunda 2009; van Rensburg 2017; Leakey 2020).

The benefits of this approach are well-documented in sub-Saharan Africa. Permaculture has been shown to improve food production, rehabilitate degraded farmland, and create new local business opportunities that can kick-start rural economies and enhance social well-being (van Rensburg 2017; Leakey 2020). Moreover, the emphasis on efficient use and conservation of resources, such as water and energy, addresses the pressing challenges of resource scarcity in the region (Tenywa and Bekunda 2009).

Challenges to the widespread adoption of permaculture in sub-Saharan Africa do exist, such as barriers to change and the need for specialist knowledge from natural and social sciences. However, the transformative potential of this approach, particularly in the context of community engagement and knowledge sharing, cannot be overstated. As the region grapples with complex, interrelated issues of food security, environmental degradation, and social inequity, the holistic and participatory nature of permaculture offers a promising pathway toward sustainable development and improved livelihoods for rural communities (Tenywa and Bekunda 2009; van Rensburg 2017; Blake et al. 2018; Leakey 2020).

8.14 Agroforestry and Agroecology: Sustainable and Resilient Agricultural Approaches in Sub-Saharan Africa

Agroforestry and agroecology have emerged as crucial approaches in promoting sustainable and resilient agricultural systems in sub-Saharan Africa. Agroforestry refers to the growing of trees and crops, sometimes with animals, in interacting combinations on the same unit of land, offering sustainable solutions to issues such as food security, environmental protection, and climate change mitigation (Nair 2013). These practices, which have been prevalent for centuries in different parts of the world, have gained scientific attention and improvements in recent decades, including the integration of multipurpose trees that produce a variety of products and services.

The traditional agroforestry practices that have been the focus of scientific research and development include tree intercropping, home gardens, shaded perennial cash crops, silvopasture, windbreaks, and tree woodlots for land rehabilitation and fallow regeneration (Nair 2007, 2014). These integrated systems have been shown to improve food and nutritional security while also mitigating environmental degradation, offering a sustainable alternative to monoculture production (Nair 2007). Agroforestry's ability to provide supportive and complementary roles across a range of landscapes and economies has been increasingly recognized, with more research and effort needed to explore its full potential.

Agroecology, on the other hand, is a holistic approach to agriculture that focuses on the interactions between various components of the agroecosystem, including plants, animals, and the environment. Rooted in traditional farming practices, agroecology emphasizes the use of locally adapted, diverse, and resilient crops and livestock, as well as the conservation and enhancement of natural resources. By leveraging the unique properties and interactions of microorganisms, agroecology can also play a crucial role in addressing abiotic stress in agriculture and promoting soil and plant health.

The integration of agroforestry and agroecological principles has the potential to transform agricultural systems in sub-Saharan Africa, promoting sustainable and resilient food production while also addressing broader environmental and social challenges.

8.14.1 Agroforestry Systems in Sub-Saharan Africa: Enhancing Sustainability and Productivity

Agroforestry, the deliberate incorporation of trees, crops, and/or livestock within a farming system, has emerged as a sustainable land management approach in sub-Saharan Africa (Nair 2013, 2014). This integrated system offers a range of benefits, including improved soil fertility, reduced erosion, and enhanced microclimate conditions for agricultural productivity (Nair 2007, 2013, 2014; Fauziyah et al. 2021).

One common agroforestry practice in the region is alley cropping, where rows of trees or shrubs are planted between crop rows. This system leverages the nitrogen-fixing capabilities of leguminous trees to enrich the soil, while also providing shade and wind protection to the crops (Koskey et al. 2021). Silvopastoral systems, which integrate trees or shrubs with livestock grazing areas, are another widely adopted agroforestry approach. The presence of trees in these systems offers shade and shelter for animals, improving their welfare and productivity, in addition to contributing to forage production and soil fertility (Nair 2013, 2014).

Windbreaks, consisting of linear hedgerows of trees or shrubs, are strategically placed to shield crops and livestock from strong winds, reducing erosion and creating more favorable microclimates (Nair 2013; Nair 2014; Kanzler et al. 2021). Home gardens, or agroforestry gardens within residential areas, also represent a small-scale agroforestry system that typically includes a diverse mix of fruit trees, vegetables, medicinal plants, and sometimes small livestock (Nair 2013; Nair 2014; Kanzler et al., 2021; Koskey et al. 2021).

The choice of tree species used in these agroforestry systems depends on various environmental, social, and economic factors. Recent development programs in the Sahel region have promoted the use of leguminous trees, such as acacia, to combat desertification and improve soil health (Koskey et al. 2021).

8.14.2 Enhancing Biodiversity Conservation and Ecosystem Services

Agroforestry, as a sustainable land management practice, plays a crucial role in promoting biodiversity conservation and enhancing ecosystem services. The integration of trees within agricultural landscapes increases habitat diversity, providing shelter, nesting sites, and food sources for a wide range of plant and animal species (Bhagwat et al. 2008). This not only supports the survival and thriving of diverse organisms but also contributes to the overall resilience and sustainability of agricultural systems (Nair 2013).

One of the key benefits of agroforestry in promoting biodiversity is the improvement of soil fertility. The leaf litter and organic matter contributed by the trees help enrich the soil, fostering beneficial soil microbial activity and providing nutrients for the growth of diverse plant species, including understory vegetation (Nair 2013). This in turn supports a wide range of insects, birds, and other organisms that rely on the thriving plant life (Baliton et al. 2020).

Agroforestry systems also contribute to water conservation and management, which are essential for sustaining biodiversity. The presence of trees in agroforestry systems helps to regulate water flows, reduce runoff, and enhance water infiltration into the soil (Nair 2013; Fauziyah et al. 2021). Their extensive root systems bind the soil, prevent erosion, and improve water-holding capacity, creating favorable conditions for moisture-loving plants and organisms, including amphibians and insects, to thrive (Bhagwat et al. 2008; Nair 2007; Baliton et al. 2020).

Furthermore, agroforestry systems play a significant role in mitigating climate change and supporting biodiversity conservation through carbon sequestration. Trees in agroforestry systems are effective at capturing and storing carbon dioxide through photosynthesis, reducing greenhouse gas emissions and contributing to the overall carbon stock.

8.14.3 The Role of Agroforestry in Enhancing Soil Health and Nutrient Cycling

Agroforestry systems have gained increasing recognition for their significant positive impacts on soil health and nutrient cycling, making them valuable tools for sustainable agriculture. This text will delve into the ways in which agroforestry contributes to improving soil health and enhancing nutrient cycling, supported by relevant citations.

One of the key contributions of agroforestry to soil health is its ability to prevent soil erosion. The presence of trees in agroforestry systems provides protective cover, reducing the impact of rainfall on the soil surface. Tree canopies intercept rainfall, dissipating its energy and preventing the direct impact of raindrops on the soil (Nair 2011). This protective effect minimizes soil compaction, maintains soil structure, and reduces surface runoff, thus preventing erosion and preserving soil integrity (Colmenares et al. 2020).

Agroforestry also enhances soil fertility and organic matter content. Trees contribute to the accumulation of organic matter through the deposition of leaf litter and the addition of woody biomass. The decomposition of leaf litter and organic residues improves soil structure, increases water-holding capacity, and promotes nutrient availability. The incorporation of organic matter into the soil creates a favorable environment for soil microorganisms and beneficial soil fauna, further enhancing nutrient cycling processes (Nair 2011; Pinho et al. 2012; Colmenares et al. 2020).

Nitrogen cycling is a critical aspect of nutrient management in agroforestry systems. Certain tree species, known as nitrogen-fixing trees, have the unique ability to capture atmospheric nitrogen and convert it into a form that is available for plant uptake. Through symbiotic associations with nitrogen-fixing bacteria, these trees fix nitrogen and release it into the soil, benefiting neighboring crops. This natural nitrogen fixation process reduces the reliance on synthetic fertilizers and contributes to the sustainability of nutrient management.

Furthermore, the presence of trees in agroforestry systems enhances nutrient cycling by facilitating the uptake and redistribution of nutrients from deeper soil horizons. The extensive root systems of trees can access nutrients that are beyond the reach of annual crops, drawing them up and depositing them on the soil surface through leaf litter and other organic matter. Some tree species, such as those with deep taproots, are particularly adept at capturing and recycling nutrients that would otherwise be lost from the system through leaching or erosion.

This agroforestry approach to soil health and nutrient cycling has been widely recognized for its potential to improve food production in a sustainable manner. Agroforestry practices can diversify farms and enhance resilience, contributing to improved food security and environmental sustainability.

8.15 The Role of Agroforestry in Climate Change Adaptation and Mitigation in Sub-Saharan Africa

Agroforestry systems in sub-Saharan Africa play a pivotal role in addressing the challenges posed by climate change, contributing to both adaptation and mitigation strategies. One of the key ways agroforestry supports climate change adaptation is through its ability to mitigate heat stress on crops and livestock (Verchot et al. 2007). The presence of trees in agroforestry systems creates a microclimate that reduces temperature extremes, mitigates heat stress, and helps maintain favorable conditions for crop growth (Verchot et al. 2007). Shade-giving trees can protect sensitive crops from excessive solar radiation, preventing heat-induced damage and improving overall crop productivity (Verchot et al. 2007).

Agroforestry systems also serve as effective windbreaks and shelterbelts, offering protection against strong winds. In regions prone to strong winds, such as coastal areas or open landscapes, the strategic placement of trees in agroforestry systems helps reduce wind speed and minimize wind-related damage to crops and soil erosion. By acting as windbreaks, trees contribute to the resilience of agricultural landscapes and help maintain the integrity of farming systems (Verchot et al. 2007; Koskey et al. 2021).

Moreover, agroforestry systems contribute to climate change mitigation by sequestering carbon dioxide from the atmosphere. Trees are efficient in capturing and storing carbon in their biomass, thereby reducing the concentration of greenhouse gases in the atmosphere. Agroforestry systems, with their combination of trees and agricultural crops or livestock, have the potential to sequester significant amounts of carbon. The diverse tree species and longer rotation cycles characteristic of agroforestry systems enhance carbon storage, making them valuable in mitigating climate change (Verchot et al. 2007; Jayne et al. 2018; Koskey et al. 2021).

The diversity and resilience of agroforestry systems make them well-suited for climate change adaptation. The presence of multiple tree species, combined with agricultural crops or livestock, provides a range of ecosystem services and reduces the vulnerability of the system to environmental stresses. Additionally, agroforestry has been shown to play a significant role in mitigating the atmospheric accumulation of greenhouse gases, further contributing to climate change mitigation efforts in the region.

8.16 Agroforestry Systems: Enhancing Livelihoods and Food Security in Sub-Saharan Africa

Agroforestry systems in sub-Saharan Africa offer a multifaceted approach to improving livelihoods and food security for farmers. These integrated systems that combine trees, crops, and sometimes livestock provide diverse sources of income and food, contributing to economic resilience and dietary diversity (Nair 2014; Leakey 2017; Koskey et al. 2021).

One of the key advantages of agroforestry is the wide range of marketable products that trees can offer. Trees in agroforestry systems can provide a variety of products such as fruits, nuts, timber, and medicinal plants, which can be sold locally or regionally to generate income for farmers (Nair 2014; Kumar et al. 2018a, b, c). For instance, fruit trees like mangoes, citrus fruits, and avocados can be harvested and sold, providing a valuable source of income (Nair 2014). Additionally, timber from selected tree species can be utilized for construction or sold as a raw material, contributing to local economies (Nair 2014; Leakey 2017).

Agroforestry systems also enhance food security by diversifying agricultural production. The integration of fruit and nut trees in these systems adds nutritional diversity to the diet, improving food security and dietary quality. Moreover, the availability of tree products within the farming system ensures a more stable and sustainable food supply throughout the year, complementing the harvest of annual crops (Nair 2014; Leakey 2017).

The diversity of agroforestry systems also creates additional income streams for farmers. In addition to the sale of tree products, agroforestry systems can provide opportunities for value-added processing, such as making jams, juices, or herbal products from fruits and medicinal plants. This value-added processing increases the economic returns from agroforestry products and supports the development of small-scale enterprises within farming communities (Nair 2014; Leakey 2017; Kumar et al. 2018a, b, c; Koskey et al. 2021).

8.17 The Role of Perennial Crops in Agroecological Systems

Perennial crops, which are plants that live for more than two years, have garnered increasing attention in the realm of agroecology due to their ability to promote sustainable and resilient agricultural practices. These crops, which include tree fruits, nuts, and sugarcane, among others, comprise a significant portion of the world's croplands, accounting for approximately 30% of the global total. However, the impacts of perennial crop cultivation on soil properties and ecosystem services have been widely variable, with some studies reporting higher soil organic carbon levels, while others have found lower or no significant changes.

This variation is likely attributable to the diverse range of climates, soil types, crop types, and management practices associated with perennial crop systems.

Nonetheless, the potential benefits of perennial crops in agroecological contexts are substantial. Perennial crops can improve soil quality by reducing erosion, increasing organic matter, and enhancing water infiltration, as well as supporting biodiversity and more resilient farming systems. For instance, the integration of perennial grasses into peanut-cotton rotations has been shown to improve soil properties, reduce pest pressure, and increase crop yields (Katsvairo et al. 2006). Similarly, the introduction of a perennial rice variety, PR23, has demonstrated comparable yields to annual rice, while also offering cost and labor savings, as well as likely benefits to soil stability and ecological sustainability (Huang et al. 2018).

While the full assessment of the ecosystem services provided by perennial cropping systems requires a more holistic analysis, the available evidence suggests that the strategic integration of these crops into agricultural landscapes can contribute to the development of more sustainable and resilient agroecological systems (Katsvairo et al. 2006; Weißhuhn et al. 2017; Huang et al. 2018; Ledo et al. 2019).

8.17.1 The Long-Term Sustainability Advantages of Perennial Crops

Perennial crops, in contrast to annual crops, have extended lifespans and can persist for multiple growing seasons, offering several advantages that contribute to the long-term sustainability of agricultural systems.

One key benefit of perennial crops is the reduced risk of soil erosion. Since perennial crops remain in the ground for several years, they provide continuous ground cover, protecting the soil from the impact of raindrops and minimizing erosion caused by wind. The extensive root systems of perennial crops also help bind the soil particles together, further enhancing soil stability and reducing erosion.

The presence of perennial crops in agricultural systems also plays a crucial role in preserving soil structure. Unlike annual crops, which require regular plowing and tillage, perennial crops can be managed with minimal soil disturbance. This reduced soil disturbance helps maintain the natural soil structure and prevents the breakdown of soil aggregates, which are essential for maintaining good soil porosity, water infiltration, and nutrient exchange. Preserving soil structure promotes better nutrient cycling, enhances water-holding capacity, and improves overall soil health and fertility (Ledo et al. 2019; Bacq-Labreuil et al. 2018).

Furthermore, the longevity of perennial crops allows them to contribute significantly to the buildup of soil organic matter over time. Perennial crops, with their extensive root systems and persistent above-ground biomass, continually add organic matter to the soil through root exudates, litterfall, and decomposition of plant residues. This organic matter serves as a source of nutrients, improves soil structure, enhances microbial activity, and increases the water-holding capacity of the soil (Katsvairo et al. 2006; Bacq-Labreuil et al. 2018; Huang et al. 2018; Ledo et al. 2019). The accumulation of organic matter in the soil not only improves soil fertility but also promotes carbon sequestration, helping mitigate climate change by reducing atmospheric carbon dioxide levels (Katsvairo et al. 2006; Bacq-Labreuil et al. 2018; Huang et al. 2018; Ledo et al. 2019).

Numerous studies have demonstrated the positive impacts of perennial crops on soil properties and long-term sustainability (Ledo et al. 2019). These benefits, combined with the reduced labor and input requirements of perennial systems, make a strong case for the widespread adoption of perennial crops as a key strategy for sustainable agriculture.

8.17.2 The Importance of Perennial Crops in Soil Health and Nutrient Cycling

Perennial crops possess extensive root systems that can penetrate deeper soil layers, offering several advantages in terms of nutrient uptake and cycling (Katsvairo et al. 2006; Nayak et al. 2020). Unlike shallow-rooted annual crops, perennial crops are equipped with roots that can explore deeper soil horizons, allowing them to access nutrients that may be otherwise unavailable to shallow-rooted plants (Ledo et al. 2019). The ability of perennial crops to tap into these deeper soil layers enhances nutrient uptake and promotes efficient nutrient cycling within agricultural systems.

By accessing nutrients from deeper soil layers, perennial crops contribute to improved soil fertility. The deep root systems of perennial crops enable them to extract nutrients that have leached below the root zone of shallow-rooted plants. These nutrients are then incorporated into the biomass of the perennial crops and returned to the soil through litterfall and decomposition, replenishing the nutrient pool and enhancing soil fertility. The cycling of nutrients facilitated by perennial crops promotes a more balanced nutrient distribution within the soil profile, reducing the risk of nutrient imbalances and improving overall nutrient availability for plant growth (Havlin and Heiniger 2020).

In addition to nutrient cycling, perennial crops play a significant role in enhancing soil health. The extensive root systems of perennial crops contribute to the accumulation of soil organic matter, as root exudates and plant residues are incorporated into the soil. The addition of organic matter improves soil structure by promoting the formation of stable soil aggregates, which enhances water infiltration, reduces erosion, and increases soil porosity. The improved soil structure resulting from the presence of perennial crops contributes to better water retention capacity, allowing the soil to hold more moisture for plant uptake during periods of water scarcity.

Research has shown that perennial grasses can improve soil quality by reducing soil erosion and nitrate leaching, increasing organic matter content, water infiltration rates, and the abundance and diversity of micro and macro flora and fauna (Katsvairo et al. 2006). Furthermore, crops grown after perennial grasses, such as cotton and peanut, are often deeper rooted, have more vigorous growth, and can better withstand pest pressure and environmental stresses, leading to higher yields. Additionally, the integration of livestock into a cropping system with perennial grasses can make more efficient use of climate and farm resources, extending the period of productive plant growth, improving economic returns, and reducing risk by diversifying the products available for sale (Katsvairo et al. 2006; Bacq-Labreuil et al. 2018; Ledo et al. 2019).

8.17.3 The Importance of Perennial Crops in Supporting Biodiversity and Habitat Creation

Perennial crops play a vital role in maintaining biodiversity and creating diverse habitats within agricultural landscapes. These crops, with their longer lifespan and complex structure, offer a range of ecological niches that support a variety of beneficial organisms, including pollinators, birds, and insects involved in pest control (Batáry et al. 2011; Beattie and Ehrlich 2014).

The flowering periods of perennial crops provide abundant nectar and pollen resources, attracting a diverse array of pollinators such as bees, butterflies, and birds. These pollinators are essential for the reproduction of many plant species, both in the perennial crop systems and the surrounding natural vegetation, contributing to the overall pollinator diversity and abundance in the landscape. Furthermore, the complex structure and longer lifespan of perennial crops offer suitable shelters and nesting sites for birds and other wildlife, some of which play a vital role in insect pest control (Deutsch et al. 2018).

In addition to supporting pollinators and birds, perennial crops also attract a wide range of beneficial insects that contribute to biological pest control. The diverse plant structures and flowering periods of these crops attract predatory insects and parasitoids that prey on or parasitize pests, helping to maintain a balanced ecosystem and reduce the need for synthetic pesticides (Ledo et al. 2019). These natural pest control services provided by perennial crop systems can have significant economic value, as demonstrated by studies estimating the worth of natural enemy suppression of soya bean aphid in soya bean production (Power 2010).

Considering the importance of pollinators, birds, and beneficial insects in maintaining ecosystem balance and providing critical ecosystem services, the role of perennial crops in supporting biodiversity and habitat creation within agricultural landscapes cannot be overstated.

8.17.4 Water Management in Agroecological Systems: The Role of Perennial Crops

Perennial crops possess distinct advantages in terms of water management and resilience to drought conditions within agroecological systems. Their deeper root systems enable them to access water stored in deeper soil layers, which can be especially beneficial during periods of limited rainfall. The extended root networks of perennial crops allow them to tap into water sources that are inaccessible to shallow-rooted annual crops, thereby enhancing their ability to withstand and recover from drought stress.

By having deeper roots, perennial crops can access water that has percolated deeper into the soil profile, providing a more consistent water supply during dry periods (Ledo et al. 2019). This capacity to access deeper soil moisture reserves enhances their resilience to drought events and reduces their reliance on surface water sources. Furthermore, the dense canopy cover of perennial crops provides shade and reduces evaporation from the soil surface, thereby reducing water loss through evapotranspiration and improving overall water-use efficiency.

In agroecological systems, the combination of perennial crops with various water management practices can further optimize water availability and conservation. Techniques such as contouring, which involves shaping the land to create ridges and furrows along the contour lines, help slow down water runoff and facilitate water infiltration, enhancing water retention within the system. Swales, which are shallow ditches constructed on the contour, capture and retain water, preventing its rapid runoff and allowing it to infiltrate into the soil. Rainwater harvesting techniques, such as the use of catchment ponds or tanks, can capture and store rainwater during periods of high precipitation, providing an additional water source during dry spells (Stroosnijder et al. 2012; Jager et al. 2020; Kumawat et al. 2020). By combining perennial crops with these water management practices, agricultural systems can become more resilient to drought and water scarcity, ensuring sustainable food production even in the face of a changing climate (Hudson 1987; Stroosnijder et al. 2012; Jager et al. 2020).

8.17.5 The Role of Perennial Crops in Climate Change Mitigation and Resilience

Perennial crops play a crucial role in climate change mitigation by actively sequestering carbon dioxide (CO_2) from the atmosphere and storing it in their biomass and soil (Ledo et al. 2019). The ability of perennial crops to continuously capture and store carbon over multiple years makes them effective tools in reducing greenhouse gas emissions and combating climate change (Morgan et al. 2010).

The above-ground biomass of perennial crops, including stems, leaves, and branches, serves as a significant carbon sink. These crops have extensive and persistent above-ground biomass, allowing for continuous carbon uptake and storage throughout their long lifespans (Ledo et al. 2019). The sequestered carbon remains locked within the perennial crop system, reducing the amount of CO_2 released into the atmosphere. Furthermore, the deep root systems of perennial crops contribute to carbon storage in the soil. As the roots of perennial crops grow deeper into the soil profile, they contribute to the accumulation of organic matter in the form of root exudates, root turnover, and root detritus, enhancing soil carbon content and promoting long-term carbon sequestration (Ledo et al. 2019).

In addition to carbon sequestration, the deep root systems of perennial crops provide other climate-related benefits. The deep roots improve soil moisture retention by accessing water stored in deeper soil layers, thereby enhancing the resilience of crops to drought conditions, while also reducing soil erosion and helping to retain carbon within the soil, preventing its loss to the atmosphere and contributing to climate resilience (Swift 2001; Morgan et al. 2010; Mayer et al. 2018; Ledo et al. 2019).

Overall, the cultivation of perennial crops contributes significantly to climate change mitigation efforts. Their ability to sequester carbon dioxide from the atmosphere and store it in both above-ground biomass and soil, along with their deep root systems, enhances the resilience of agroecosystems to the impacts of climate change (Huang et al. 2018; Mayer et al. 2018; Ledo et al. 2019; Liu et al. 2024).

8.17.6 The Economic and Food Security Benefits of Perennial Crops

Perennial crops offer significant economic advantages to farmers, primarily due to their extended lifespan and potential for multiple harvests, which provide a more consistent and reliable income stream compared to annual crops that require replanting each growing season. The longer lifespan of perennial crops allows farmers to reap the benefits of their investments over an extended period, reducing the need for yearly planting efforts and associated costs (Ledo et al. 2019). Farmers who adopt perennial crops can shift to a more long-term perspective and benefit from the continuous production and marketability of these crops, leading to stable incomes and reduced income volatility (Glover et al. 2010; Huang et al. 2018).

Many perennial crops also offer diverse marketable products, such as fruits, nuts, oils, fibers, and medicinal plants, adding to their economic potential. This diversification of income streams enhances the financial resilience of farmers and provides opportunities for value-added processing and marketing (Huang et al. 2018).

The inclusion of perennial crops in agroecological systems can also contribute to improved food security. The longer lifespan and resilience of perennial crops make them more suitable for areas prone to climate variability and extreme weather events, allowing them to withstand and recover from disturbances better than annual crops. Perennial crops can provide a more stable and reliable food supply, reducing the vulnerability of farmers and communities to food shortages in times of climatic uncertainty (Katsvairo et al. 2006; Glover et al. 2010; Huang et al. 2018; Ledo et al. 2019).

Perennial grains, in particular, hold promise for marginal landscapes or areas with limited resources where annual versions may struggle. Perennial crops have also been shown to have positive effects on soil organic carbon, though the results from the literature are inconclusive, potentially due to the diverse nature of published studies (Ledo et al. 2019).

8.18 Grassland Management Strategies in Sub-Saharan Africa

Grassland management in Sub-Saharan Africa is a critical component of agroecological practices, as these expansive grassland regions support a diverse array of agricultural and pastoral systems that are vital to the livelihoods and food (Kindomihou 2020). Across this region, grasslands face increasing pressure due to population growth and land degradation, which has led to only marginal increases in cereal production over the past two decades, with most of this growth attributable to area expansion rather than yield. Women play a crucial role in grassland management, contributing up to 70% of household food production and heading approximately a third of rural households in sub-Saharan Africa.

The availability of feed significantly impact livestock well-being, productivity, and reproduction. However, many small livestock keepers and pastoralists still face challenges in maintaining basic biosecurity standards for animal health. While progress is being made in improving livestock feed, the lack of reasonably priced, high-quality feed remains a significant obstacle to increasing livestock productivity and meeting the growing demand for livestock products in the region (Bayu 2020; Kindomihou 2020; Panel 2020).

Adopting better agronomic management practices, such as those shown to increase the climate resilience of maize to drought in Tanzania, could be a key strategy for enhancing grassland-based production systems in sub-Saharan Africa (Xiong and Tarnavsky 2020). Such practices could help smallholder farmers, who are crucial for production growth and poverty alleviation, to adapt to the challenges posed by climate change, population growth, and land degradation (Jayne et al. 2018; Bayu 2020; Panel 2020; Xiong and Tarnavsky 2020).

8.18.1 Rotational Grazing: Sustaining Grassland Ecosystems

Agroecology, a field that emphasizes the integration of ecological principles into agricultural practices, places significant emphasis on sustainable grazing strategies that prioritize the well-being and integrity of grassland ecosystems. One such strategy is the implementation of rotational grazing systems, also known as managed intensive grazing or planned grazing (Dong et al. 2020; de Otálora et al. 2021).

In rotational grazing systems, livestock are regularly moved between different grazing areas or paddocks, allowing for rest and recovery periods for the grasses and vegetation. This approach offers numerous environmental benefits. By avoiding continuous grazing in a single area, these systems help prevent overgrazing and give grasses the opportunity to regrow and replenish their energy reserves. This promotes the overall health of the grassland ecosystem, as it allows for proper root development, increased photosynthetic capacity, and improved nutrient cycling (Dong et al. 2020; de Otálora et al. 2021; Cozim-Melges et al. 2024).

Resting periods in rotational grazing systems also provide opportunities for biodiversity to flourish. As the grasses and vegetation recover, they can provide habitat and food sources for a wide range of beneficial organisms, including insects, birds, and small mammals. This, in turn, promotes biodiversity within the grassland ecosystem and supports ecological balance (Delaby et al. 2020; Dong et al. 2020).

Furthermore, rotational grazing systems contribute to the enhancement of soil fertility. Resting periods allow for the accumulation of organic matter in the soil, as plant residues decompose and contribute to soil organic carbon. This organic matter improves soil structure, increases water-holding capacity, and enhances nutrient cycling. The movement of livestock between grazing areas also distributes nutrient-rich manure more evenly across the land, ensuring better nutrient utilization by plants (Bohnert and Stephenson 2016; Delaby et al. 2020; Dong et al. 2020).

Regenerative rotational grazing management, as demonstrated in a study conducted in the Basque Country of northern Spain, has been shown to achieve higher grass production and carbon storage compared to traditional grazing practices (de Otálora et al. 2021). Similarly, research has highlighted the ability of rotational grazing to maintain or improve rangeland health and ecological function in a way that is economically viable, through the use of supplementation strategies and complimentary forages.

8.18.2 The Importance of Preserving Biodiversity in Sub-Saharan African Grasslands

Sub-Saharan African grasslands are renowned for their remarkable biodiversity, hosting a diverse array of plant and animal species (Mokany et al. 2020). Agroecological approaches recognize the crucial importance of preserving and enhancing this biodiversity as a fundamental aspect of sustainable agriculture. By promoting diverse plant communities and encouraging the presence of native grasses and forbs, agroecology aims to maintain the ecological integrity and resilience of these grassland ecosystems.

One key strategy employed in agroecology is the use of diverse plant communities in grasslands. This involves incorporating a variety of plant species with different growth habits, root structures, and flowering times (Cozim-Melges et al. 2024). The presence of diverse plants not only increases overall species richness but also provides a range of resources and habitats for different organisms (Thrupp 2000). For example, native grasses and forbs serve as food sources for herbivores and provide shelter and nesting sites for birds and small mammals (Batáry et al. 2011; Thrupp 2000).

A grassland ecosystem in East Africa

Maintaining a diverse plant community in grasslands is particularly important for supporting pollinators and beneficial insects. Pollinators, such as bees, butterflies, and beetles, play a crucial role in plant reproduction and the production of fruits and seeds, as well as contributing to the maintenance of plant genetic diversity. By promoting diverse plant species and providing suitable habitats, agroecological approaches enhance pollinator abundance and diversity, which in turn benefits crop pollination and ecosystem stability (Thrupp 2000; Batáry et al. 2011; Yang and Chen 2021; Birkhofer et al. 2022).

Furthermore, diverse plant communities in grasslands support a variety of beneficial insects that contribute to pest control and natural balance within agroecosystems. These beneficial insects include predators, parasitoids, and decomposers that help regulate populations of pests and contribute to the overall health and resilience of the ecosystem.

In summary, agroecology recognizes the rich biodiversity of sub-Saharan African grasslands as a fundamental aspect of sustainable agriculture. By promoting diverse plant communities and enhancing the presence of native grasses and forbs, agroecological approaches aim to maintain the ecological integrity and resilience of these grassland ecosystems, supporting a wide range of organisms and the essential ecosystem services they provide (Edwards et al. 1993; Forister et al. 2019; Koskey et al. 2021; Cozim-Melges et al. 2024).

8.18.3 Soil Conservation and Restoration: Agroecological Approaches in Grassland Management

Grassland management practices in agroecology place a strong emphasis on soil conservation and restoration to maintain the health and productivity of grassland ecosystems. By implementing various techniques, such as contour plowing, grassed waterways, and terracing, agroecological approaches aim to prevent soil erosion on sloping grasslands and promote sustainable land use (Wade et al. 2008).

Contour plowing is a technique commonly used in agroecology to reduce soil erosion on hilly or sloping grasslands. It involves plowing along the contour lines of the land, creating furrows that follow the natural contours of the terrain. This method helps slow down the flow of water and minimizes the risk of runoff, allowing more water to infiltrate into the soil and reducing erosion. Grassed waterways are another important tool for soil conservation in agroecological grassland management. They are designed to channel water runoff, particularly during heavy rain events, through vegetated channels. The presence of grasses and other vegetation in these waterways helps to trap sediment and stabilize the soil, reducing erosion and the transport of sediments downstream (Sanam et al. 2022). Terracing is a practice commonly employed on steeply sloping grasslands to create level or gently sloping platforms. Terraces help to break the flow of water down the slope, reducing erosion and promoting water infiltration. By reducing the speed and force of runoff, terraces help to prevent soil erosion and create more favorable conditions for grassland vegetation (Hudson 1987).

In addition to soil conservation, agroecology also focuses on restoring degraded grasslands to improve their soil health and productivity. Reseeding with native grass species is a common practice in grassland restoration efforts. Native grasses are adapted to the local environmental conditions, have deep root systems, and can provide better soil stabilization and nutrient cycling compared to other species (Wade et al. 2008; Cozim-Melges et al. 2024). Proper management of grazing enterprises can also support the regeneration of soil function, improving essential ecosystem services such as carbon sequestration, water infiltration, soil fertility, nutrient cycling, and biodiversity (Teague 2018). Collectively, conservation agriculture managed regeneratively can support ecologically healthy, resilient agroecosystems and enhance watershed function (Birkhofer et al. 2022; Teague 2018).

8.18.4 The Role of Agroecological Grassland Management Practices in Carbon Sequestration

Grasslands play a crucial role in climate change mitigation through their capacity to sequester and store carbon dioxide (CO_2) from the atmosphere. Implementing effective agroecological management practices in grasslands can enhance their ability to sequester and store carbon, thereby contributing to climate change mitigation efforts (Dlamini et al. 2016; Amelung et al. 2020).

Rotational grazing is a key practice in agroecological grassland management that involves periodically moving livestock between different grazing areas. This practice allows for rest and recovery of grasses, promoting their growth and increasing carbon sequestration in the soil (Morgan et al. 2010). Rotational grazing reduces overgrazing and allows for the accumulation of organic matter, which contributes to soil carbon storage (Dlamini et al. 2016). Maintaining vegetative cover is another important aspect of agroecological grassland management that enhances carbon sequestration. By ensuring that grasslands have a continuous cover of vegetation, such as grasses and forbs, soil erosion is minimized, and organic matter inputs to the soil are increased, leading to the sequestration of carbon and mitigation of climate change (Dlamini et al. 2016; Liu et al. 2024).

Integrating trees or agroforestry systems with grasslands can further enhance carbon storage and ecosystem resilience. Trees in agroforestry systems contribute to carbon sequestration through the uptake of CO_2 during photosynthesis and the accumulation of biomass in their above-ground and belowground parts. The presence of trees in grassland areas can enhance carbon sequestration by increasing the overall vegetation cover and providing additional organic matter inputs to the soil (Morgan et al. 2010; Zomer et al. 2017; Mayer et al. 2018; Liu et al. 2024).

The combined effect of rotational grazing, maintaining vegetative cover, and integrating trees or agroforestry systems with grasslands can lead to significant carbon sequestration benefits. These practices enhance the capacity of grassland ecosystems to sequester and store carbon, contributing to climate change mitigation efforts (Morgan et al. 2010; Zomer et al. 2017; Mayer et al. 2018; Liu et al. 2024).

8.18.5 Sustainable Pastoralism in Agroecological Grassland Management

Grasslands in sub-Saharan Africa are not merely valuable ecosystems but also serve as crucial resources for pastoralist communities, supporting their livelihoods and cultural traditions. Agroecological approaches to grassland management recognize the importance of integrating social and economic aspects into sustainable pastoralism practices, aiming to create a balanced and mutually beneficial relationship between humans and the environment (Koocheki and Gliessman 2005; Liu et al. 2019; Kindomihou 2020). Sustainable pastoralism in agroecological grassland management involves implementing practices that support the social and economic needs of pastoralist communities while ensuring the long-term health and productivity of grasslands.

Pastoral nomadism, a complex set of practices and knowledge, has permitted the long-term maintenance of a sophisticated "triangle of sustainability" that includes plants, animals, and people (Koocheki and Gliessman 2005). Pastoralism is presented as the integration of cropping systems, domestic animal management, natural resource conservation, and sociocultural organization, making it a viable and practical alternative for sustaining the landscape and the people who occupy it in strongly water-limited parts of the world (Koocheki and Gliessman 2005). In response to the challenges faced by pastoralist communities, such as land tenure issues, access to water resources, and conflicts over grazing rights, pastoralists are creating new ways to manage rangelands through conservancies and community-based institutions on state, common, and private land (Koocheki and Gliessman 2005; Reid et al. 2014).

Agroecological management of grasslands also recognizes the need to address the economic aspects of pastoralism. This involves exploring alternative income-generation opportunities for pastoralist communities, diversifying their sources of livelihood, and improving their economic resilience (Panel 2020). Agroecology may promote the development of value-added products from livestock, such as dairy processing or wool production, which can enhance the economic viability of pastoralism (Koocheki and Gliessman 2005; Reid et al. 2014; Tadesse et al. 2015; Panel 2020). By engaging with pastoralist communities, empowering them with knowledge and resources, and promoting participatory decision-making processes, agroecology aims to support sustainable pastoralism while safeguarding the ecological balance of grassland ecosystems.

8.18.6 Participatory Approaches and Local Knowledge in Grassland Management: An Agroecological Perspective

Agroecology recognizes the crucial role of local knowledge and active community participation in the sustainable management of grasslands (Altieri 2002; Lacombe et al. 2018; Pagliarino et al. 2020). It emphasizes the value of integrating traditional ecological knowledge held by pastoralists, farmers, and indigenous communities into decision-making processes and management strategies. By involving local stakeholders, agroecology promotes context-specific and sustainable approaches to grassland stewardship that are rooted in the expertise and lived experiences of those who intimately interact with these ecosystems.

Participatory approaches are central to agroecological grassland management. These approaches involve actively engaging local communities in all stages of decision-making, planning, and implementation (Lacombe et al. 2018). Through participatory processes, communities are empowered to contribute their knowledge, perspectives, and priorities, shaping more inclusive and effective grassland management strategies (Lacombe et al. 2018). This collaboration fosters a sense of ownership and responsibility among community members, leading to improved long-term outcomes and the sustainability of grassland ecosystems (Altieri 2002; Chirenje et al. 2013; Lacombe et al. 2018).

The integration of local knowledge is a key principle of agroecology, as it encompasses the accumulated wisdom, observations, and practices developed by generations of pastoralists, farmers, and indigenous peoples living in close relationship with grasslands. This knowledge offers insights into the complex dynamics of grassland ecosystems, including traditional grazing practices, mobility patterns, and resource management techniques (Altieri 2002; Suárez-Torres et al. 2017). Integrating local knowledge into grassland management allows for the adaptation of strategies to specific ecological, social, and cultural contexts, ensuring more effective and sustainable outcomes (Altieri 2002; Bisong et al. 2017).

By valuing and incorporating local knowledge, agroecology recognizes the inherent expertise of local communities in grassland management (Wisner 2010). This aligns with the ecosystem-based perspective of agroecology, which emphasizes the importance of understanding the complex web of biological, physical, and chemical interactions that underpin the productivity and resilience of agricultural systems (Gliessman 1995). Agroecological approaches are particularly useful for agrarian systems, as they provide holistic information about the physical functioning and spatial/temporal differences of these systems, including the links between natural resources and consumer markets.

8.19 The Role of Seeds and Planting Materials in Agroecological Practices in Sub-Saharan Africa

Across sub-Saharan Africa, the management and exchange of seeds and planting materials play a crucial role in the success and sustainability of agroecological practices. Robust seed systems, access to appropriate seeds and planting materials, and the preservation of traditional seed exchange networks are all essential components for enabling small-scale farmers to adopt and maintain agroecological approaches.

The informal seed sector, where farmers save, exchange, and market their own seeds, remains the primary source of seed acquisition for the majority of smallholder farmers in the region. Estimates indicate that up to 90% of the seed used by smallholder farmers in sub-Saharan Africa is sourced through these informal channels, with a significant portion (over 50%) coming from local markets. This informal seed system is particularly adept at providing farmers with the seeds of traditional crop varieties well-suited to local growing conditions, but can struggle to effectively disseminate newly released improved cultivars (Grisley 1993).

The social and cultural dimensions of traditional seed exchange networks also contribute to the resilience of smallholder farming systems. These exchange systems cement community and family bonds, provide a safety net in times of harvest failure, and reinforce local food sovereignty. However, the value and role of these informal systems are often overlooked in development efforts that have tended to prioritize strengthening formal seed supply chains.

To enhance the sustainability of agroecological practices in sub-Saharan Africa, a more holistic approach is needed that supports both the formal and informal seed sectors. Interventions should focus on improving farmer access to a diverse portfolio of quality seeds and planting materials, while also recognizing and preserving the important social and cultural functions of traditional seed exchange networks (Grisley 1993; McGuire and Sperling 2016; van Niekerk and Wynberg 2017; Sperling et al. 2020).

8.20 Agrobiodiversity Conservation: Preserving Genetic Diversity for Sustainable Agriculture

Agroecology places a strong emphasis on the use and conservation of diverse and locally adapted crop varieties, recognizing the value of traditional and heirloom seeds that have been passed down through generations and adapted to local agroecological conditions over time (Thrupp 2000; Altieri 1999; Gill et al. 2014). By promoting the use of these seeds, agroecology aims to preserve biodiversity, enhance ecosystem resilience, and support sustainable farming systems (Thrupp 2000).

Traditional and heirloom seeds exhibit genetic diversity, providing a broad range of traits that enable crops to adapt and thrive in various environmental conditions (Thrupp 2000; Nodari and Guerra 2015). This genetic diversity can enhance the resilience of crops to pests, diseases, and climate variability (Altieri 1999). These seeds have often evolved through farmer selection, developing natural defense mechanisms that help protect crops without the need for synthetic pesticides, thereby reducing the dependence on chemical inputs (Altieri 1999; Gill et al. 2014; Nodari and Guerra 2015).

Moreover, traditional and heirloom seeds are often well-adapted to local climatic conditions and can exhibit greater tolerance to drought, heat, or other climatic stressors. This adaptability is crucial in the face of climate change, enabling farmers to maintain crop productivity even under changing environmental conditions (Nodari and Guerra 2015). By utilizing locally adapted seeds, farmers can reduce the risk of crop failure and improve the resilience of their farming systems (Thrupp 2000; Gill et al. 2014; Nodari and Guerra 2015).

Agroecology recognizes the importance of farmers' roles as custodians of seed diversity and promotes the conservation of traditional and heirloom seeds to ensure their availability for future generations. This conservation can take various forms, including on-farm seed saving and seed exchange, which help maintain the genetic diversity of these valuable resources (Altieri 1999; Thrupp 2000; Gill et al. 2014; Nodari and Guerra 2015).

8.21 The Vital Role of Farmer-Managed Seed Systems in Agroecology

Farmer-managed seed systems play a crucial role in agroecology by empowering farmers to take control of their seed resources. These systems prioritize the knowledge, preferences, and rights of farmers, enabling them to save, select, and exchange seeds according to their own needs and local conditions. By actively engaging farmers in seed management, agroecology promotes seed diversity, adaptation, and resilience (Thrupp 2000; Sethuraman et al. 2021; Cozim-Melges et al. 2024).

One of the key features of farmer-managed seed systems is their emphasis on local knowledge. Farmers possess invaluable knowledge about their local agroecological conditions, including the performance of different crop varieties in specific environments. This knowledge is crucial for selecting and saving seeds that are well-adapted to local conditions, such as soil types, rainfall patterns, and pest pressures. Incorporating local knowledge into seed management can enhance the resilience and productivity of crops (Thrupp 2000; Cozim-Melges et al. 2024).

Participatory plant breeding programs are an essential component of farmer-managed seed systems, involving collaboration between farmers, plant breeders, and other stakeholders to develop new crop varieties that meet the specific needs and preferences of farmers. This approach not only leads to the development of locally adapted varieties but also strengthens farmers' capacity to manage their own seed resources (Almekinders and Louwaars 2002; McGuire and Sperling 2016). Community seed banks are another integral part of farmer-managed seed systems, serving as repositories of locally adapted seed varieties and promoting the exchange and sharing of seeds among farmers (Thrupp 2000; Almekinders and Louwaars 2002; McGuire and Sperling 2016; van Niekerk and Wynberg 2017). Farmer-managed seed systems are essential for maintaining agrobiodiversity and food security, as they empower small-scale farmers to play a central role in managing their seed resources (Thrupp 2000; Almekinders and Louwaars 2002; McGuire and Sperling 2016; van Niekerk and Wynberg 2017).

8.22 Expanding Access to Organic and Non-GMO Seeds in Sub-Saharan Africa

The growing demand for organic and non-GMO agricultural products in sub-Saharan Africa has placed a strong emphasis on agroecological practices, which prioritize ecological balance, biodiversity conservation, and the use of natural inputs (Tung 2018). To align their production with these principles, farmers often seek access to certified organic and non-GMO seeds. To meet this demand, various organizations and seed companies are actively working to provide farmers in the region with access to these specialized seed varieties (Gaffney et al. 2016).

Certified organic seed production involves adhering to strict organic standards throughout the seed production cycle, from seed selection and field management to processing and storage. This ensures that the seeds are produced using organic practices and are free from synthetic chemicals or genetically modified traits. Organic certification bodies, such as the International Federation of Organic Agriculture Movements (IFOAM), provide guidelines and certification processes for organic seed production, guaranteeing the integrity of organic seeds. Similarly, non-GMO seed production focuses on maintaining genetic purity and avoiding genetic modification. Seed companies and organizations engage in meticulous seed production and testing processes to ensure that their seeds are free from genetically modified traits, following internationally recognized protocols and standards for non-GMO seed production (Tung 2018).

However, the bean seed production and distribution sector in sub-Saharan Africa has faced challenges in supplying seed of newly released cultivars to small-scale farmers, as the informal, small farmer seed sector is efficient with regard to currently used cultivars but unable to accommodate the dissemination of new varieties (Grisley 1993). While investments have primarily focused on strengthening the formal sector, the informal sector remains the core for seed acquisition, with 90.2% of farmers accessing their seed from informal systems (McGuire and Sperling 2016).

This suggests the need for targeted interventions to render both the formal and informal seed sectors more responsive to the needs of smallholder farmers.

Linking formal and farmers' seed systems and improving the latter may in many cases be a more effective strategy to improve national and local seed supply than aiming solely at improving the formal seed sector (Almekinders and Louwaars 2002). The importance of farmers' seed systems merits closer attention at the policy level and in technical assistance projects, as analysis of the strengths and weaknesses of both systems shows important complementarity that offers multiple opportunities for improving the effectiveness of both (Almekinders and Louwaars 2002; Hampton et al. 2016).

8.23 The Role of Tree Seedling Providers in Promoting Agroforestry Practices

Agroforestry, as an integral component of agroecological systems, involves the intentional integration of trees with agricultural crops or livestock. The successful implementation of agroforestry practices relies on the availability of suitable tree seedlings for planting. Numerous organizations and nurseries are dedicated to producing and distributing high-quality tree seedlings, particularly those of indigenous and multipurpose tree species, recognizing the importance of tree diversity and functionality within agroforestry systems (Nair 2013, 2014).

Seedlings ready for transplanting in agroforestry systems

These organizations and nurseries play a crucial role in providing farmers with access to diverse and appropriate tree seedlings for agroforestry. They understand the various benefits that different tree species can bring to agroecological practices, such as nitrogen fixation, soil improvement, shade provision, fodder production, timber, or fruit yield (Baliton et al. 2020). Consequently, they focus on producing tree seedlings of species that are well-suited to the local agroecological conditions and can fulfill multiple functions within the agroforestry system (Nair 2007, 2014; Pinho et al. 2012).

The production of high-quality tree seedlings involves several stages, including seed collection, germination, nursery management, and seedling care (Koskey et al. 2021). Organizations and nurseries follow established protocols to ensure that the seedlings are healthy, vigorous, and adapted to the local environment. They employ techniques such as seed scarification, pre-treatment, and proper nursery management practices to enhance seed germination and seedling growth (Nair 2014).

To support agroforestry practices, these organizations and nurseries prioritize the cultivation of indigenous tree species. Indigenous trees are well-adapted to the local climate, soil conditions, and pests, making them resilient and suitable for agroecological systems. Furthermore, they contribute to the conservation of native biodiversity and the preservation of traditional knowledge associated with these tree species. In addition to indigenous tree species, these providers also focus on producing seedlings of multipurpose tree species that can fulfill various functions within the agroforestry system (Pinho et al. 2012; Nair 2013, 2014; Koskey et al. 2021).

The crucial role of tree seedling providers in promoting agroforestry practices is further underscored by evidence from the Sahel region of Africa, where farmer-managed agroforestry has resulted in a significant increase in food production, tree diversity, and the greening of the landscape (Koskey et al. 2021). The use of leguminous tree species, such as acacia, has been a key component of these agroforestry systems, highlighting the importance of access to appropriate tree seedlings.

8.23.1 Agroecological Seed Multiplication and Distribution Networks

Seed multiplication and distribution networks are crucial components of the agroecological seed system, ensuring the availability and accessibility of seeds suitable for agroecological farming practices. Local seed producers, cooperatives, and seed companies play a vital role in these networks, contributing to the production and dissemination of quality seeds that align with the principles of agroecology (Almekinders and Louwaars 2002).

Local seed producers, including individual farmers or community-based seed enterprises, are essential in seed multiplication efforts. They engage in seed production activities using agroecological methods, which include maintaining healthy and diverse populations of crop plants, minimizing the use of synthetic inputs, and preserving traditional farming practices (Almekinders and Louwaars 2002). By employing these practices, local seed producers contribute to the production of seeds that are well-suited to the local agroecological conditions and farming systems.

The importance of the farmers' seed systems in a functional national seed sector has been recognized, as aiming for a formal seed sector to supply 100% of the seed for planting is only realistic for a small number of crops and in few countries (Almekinders and Louwaars 2002; McGuire and Sperling 2016; Sperling et al. 2020). Cooperatives and seed companies also play significant roles in seed multiplication and distribution networks for agroecological seeds. These organizations provide the infrastructure and resources necessary for large-scale seed multiplication and commercialization. They work closely with local seed producers to ensure the production of high-quality seeds that meet agroecological standards and facilitate the dissemination of these seeds to farmers through established marketing and distribution channels (Almekinders and Louwaars 2002; Sperling et al. 2020).

In agroecological seed multiplication and distribution networks, the emphasis is on diversity and adaptation. Local seed producers, cooperatives, and seed companies recognize the importance of maintaining and enhancing the genetic diversity of crops, as it contributes to the resilience and sustainability of agricultural systems (Almekinders and Louwaars 2002; McGuire and Sperling 2016; Sperling et al. 2020). The informal seed sector, including seed traders, remains the core for seed acquisition, especially in Africa, with farmers accessing 90% of their seed from informal systems (Almekinders and Louwaars 2002; McGuire and Sperling 2016; Sperling et al. 2020).

References

Abdulai A, Hazell P (1996) The role of agriculture in sustainable economic development in Africa. J Sustain Agric 7(2–3):101–119. https://doi.org/10.1300/j064v07n02_10

Allen T, Prosperi P, Cogill B, Flichman G (2014) Agricultural biodiversity, social-ecological systems and sustainable diets. Proc Nutr Soc 73(4):498–508. https://doi.org/10.1017/s002966511400069x

Almekinders C, Louwaars N (2002) The importance of the farmers' seed systems in a functional national seed sector. J New Seeds 4(1–2):15–33. https://doi.org/10.1300/j153v04n01_02

Altieri MA (1999) The ecological role of biodiversity in agroecosystems. Agric Ecosyst Environ 74(1–3):19–31. https://doi.org/10.1016/s0167-8809(99)00028-6

Altieri MA (2002) Agroecology: the science of natural resource management for poor farmers in marginal environments. Agric Ecosyst Environ 93(1–3):1–24. https://doi.org/10.1016/s0167-8809(02)00085-3

Altieri MA, Nicholls CI (2013) The adaptation and mitigation potential of traditional agriculture in a changing climate. Clim Chang 140(1):33–45. https://doi.org/10.1007/s10584-013-0909-y

Altieri MA, Nicholls CI, Henao A, Lana M (2015) Agroecology and the design of climate change-resilient farming systems. Agron Sustain Dev 35(3):869–890. https://doi.org/10.1007/s13593-015-0285-2

Amelung W, Bossio D, Vries WD, Kögel-Knabner I, Lehmann J, Amundson R et al (2020) Towards a global-scale soil climate mitigation strategy. Nat Commun 11(1):5427. https://doi.org/10.1038/s41467-020-18887-7

Andrew J, Kelton J, Mosjidis JA (2012, February 29) Utilization of Sunn Hemp for cover crops and weed control in temperate climates. Weed Control. https://doi.org/10.5772/36601

Ansari M, Ravisankar N, Ansari MH, Babu S, Layek J, Panwar AS (2023) Integrating conservation agriculture with intensive crop diversification in the maize-based organic system: impact on sustaining food and nutritional security. Nutr Sustain Diets 10:1137247. https://doi.org/10.3389/fnut.2023.1137247

Arriaga FJ, Guzmán J, Lowery B (2017) Conventional agricultural production systems and soil functions. In: Soil health and intensification of agroecosystems, pp 109–125. https://doi.org/10.1016/b978-0-12-805317-1.00005-1

Asrat GA (2018) Mapping soil terrain resources and descriptions of agro-ecological zone in Dawuro and Gamo Gofa zones in south-western Ethiopia. Soil Sci Environ Manag J 9(10):164–179. https://doi.org/10.5897/jssem2018.0695

Ausari DK, Singh B, Aakash, Kumawat R, Gehlot Y (2020) GIS based mapping of soil fertility status of Tehsil Jobat, District Alirajpur, Madhya Pradesh, India. Int J Curr Microbiol App Sci 9(10):60–69. https://doi.org/10.20546/ijcmas.2020.910.009

Bacq-Labreuil A, Crawford J, Mooney SJ, Neal AL, Akkari E, McAuliffe C et al (2018) Effects of cropping systems upon the three-dimensional architecture of soil systems are modulated by texture. Geoderma 332:73–83. https://doi.org/10.1016/j.geoderma.2018.07.002

Baliton R, Landicho L, Cabahug RE, Paelmo RF, Laruan K, Rodriguez R et al (2020) Ecological services of agroforestry systems in selected upland farming communities in the Philippines. Biodivers J Biol Divers 21(2). https://doi.org/10.13057/biodiv/d210237

Barman S, Das S, Bhattacharya SS (2019) The prospects of bio-fertilizer technology for productive and sustainable agricultural growth. In: New and future developments in microbial biotechnology and bioengineering. Elsevier, pp 233–253. https://doi.org/10.1016/b978-0-444-64191-5.00017-1

Barrios E, Gemmill-Herren B, Bicksler A, Siliprandi E, Brathwaite R, Moller S et al (2020) The 10 elements of agroecology: enabling transitions towards sustainable agriculture and food systems through visual narratives. Ecosyst People 16(1):230–247. https://doi.org/10.1080/26395916.2020.1808705

Barzman M, Bàrberi P, Birch ANE, Boonekamp P, Dachbrodt-Saaydeh S, Graf B et al (2015) Eight principles of integrated pest management. Agron Sustain Dev 35:1199–1215. https://doi.org/10.1007/s13593-015-0327-9

Batáry P, Báldi A, Kleijn D, Tscharntke T (2011) Landscape-moderated biodiversity effects of agri-environmental management: a meta-analysis. Proc R Soc B Biol Sci 278(1713):1894–1902. https://doi.org/10.1098/rspb.2010.1923

Bayu EK (2020) Determinant variables for women's participation in soil and water conservation practices in north western Ethiopia: The case of Shebel Berenta District (Woreda), East Gojjam Zone, Amhara National Regional State. Air Soil Water Res 13:1178622120942199. https://doi.org/10.1177/1178622120942199

Beattie AJ, Ehrlich PR (2014) Industries depend on biodiversity too. Nature 509(7502):563–563. https://doi.org/10.1038/509563d

Bhagwat SA, Willis KJ, Birks HJB, Whittaker RJ (2008) Agroforestry: a refuge for tropical biodiversity? Trends Ecol Evol 23(5):261–267. https://doi.org/10.1016/j.tree.2008.01.005

Birkhofer K, Baulechner D, Diekötter T, Zaitsev A, Wolters V (2022) Fertilization rapidly alters the feeding activity of grassland soil mesofauna independent of management history. Front Ecol Evol 10:864470. https://doi.org/10.3389/fevo.2022.864470

Bisong TL, Ogbonna KI, Kyari IU (2017) Effect of community participation in forest conservation in Ikom Agricultural Zone of Cross River State. Glob J Agric Sci 16(1):31–35. https://doi.org/10.4314/gjass.v16i1.4

Blake WH, Rabinovich A, Wynants M, Kelly C, Nasseri M, Ngondya I et al (2018) Soil erosion in East Africa: an interdisciplinary approach to realising pastoral land management change. Environ Res Lett 13(12):124014. https://doi.org/10.1088/1748-9326/aaea8b

Blevins RL, Frye WW (1993) Conservation tillage: an ecological approach to soil management, pp 33–78. https://doi.org/10.1016/s0065-2113(08)60590-8

Bohnert DW, Stephenson MB (2016) Supplementation and sustainable grazing systems. J Anim Sci 94(suppl_6):15–25. https://doi.org/10.2527/jas.2016-0520

Bolinder MA, Crotty F, Elsen A, Frac M, Kismányoky T, Lipiec J et al (2020) The effect of crop residues, cover crops, manures and nitrogen fertilization on soil organic carbon changes in agroecosystems: a synthesis of reviews. Mitig Adapt Strateg Glob Chang 25:929–952. https://doi.org/10.1007/s11027-020-09916-3

Carrasco G, Urrestarazu M (2010) Green chemistry in protected horticulture: the use of peroxyacetic acid as a sustainable strategy. Int J Mol Sci 11(5):1999–2009. https://doi.org/10.3390/ijms11051999

Chenu C, Angers DA, Barré P, Derrien D, Arrouays D, Balesdent J (2019) Increasing organic stocks in agricultural soils: knowledge gaps and potential innovations. Soil Tillage Res 188:41–52. https://doi.org/10.1016/j.still.2018.04.011

Chirenje LI, Giliba RA, Musamba EB (2013) Local communities' participation in decision-making processes through planning and budgeting in African countries. Chin J Populat Resourc Environ 11(1):10–16. https://doi.org/10.1080/10042857.2013.777198

Chirinda N, Arenas L, Loaiza S, Trujillo C, Katto M, Chaparro P et al (2017) Novel technological and management options for accelerating transformational changes in rice and livestock systems. Sustain For 9(11):1891. https://doi.org/10.3390/su9111891

Cisse EHI, Ndiogou BA, Tigampo S, Thiam A, Azilinon D (2023) Optimization of a solar chimney with a horizontal absorber for building ventilation: a case study. Fluid Dyn Mater Process 19(4):901–910. https://doi.org/10.32604/fdmp.2022.021973

Colmenares OM, Brindis RC, Verduzco CV, Grajales MP, Gómez MU (2020) Horticultural agroforestry systems recommended for climate change adaptation: a review. Agric Rev 41(1):14–24. https://doi.org/10.18805/ag.r-133

Cozim-Melges F, Ripoll-Bosch R, Veen GF, Oggiano P, Bianchi FJ, van der Putten WH, van Zanten HH (2024) Farming practices to enhance biodiversity across biomes: a systematic review. NPJ Biodivers 3(1):1. https://doi.org/10.1038/s44185-023-00034-2

Delaby L, Finn JA, Grange G, Horan B (2020) Pasture-based dairy systems in temperate lowlands: challenges and opportunities for the future. Front Sustain Food Syst 4:543587. https://doi.org/10.3389/fsufs.2020.543587

Delgado JA, Short NM Jr, Roberts DP, Vandenberg B (2019) Big data analysis for sustainable agriculture on a geospatial cloud framework. Front Sustain Food Syst 3:54. https://doi.org/10.3389/fsufs.2019.00054

Deutsch CA, Tewksbury JJ, Tigchelaar M, Battisti DS, Merrill SC, Huey RB, Naylor RL (2018) Increase in crop losses to insect pests in a warming climate. Science 361(6405):916–919. https://doi.org/10.1126/science.aat3466

Dlamini P, Chivenge P, Chaplot V (2016) Overgrazing decreases soil organic carbon stocks the most under dry climates and low soil pH: a meta-analysis shows. Agric Ecosyst Environ 221:258–269. https://doi.org/10.1016/j.agee.2016.01.026

Dong S, Shang Z, Gao J, Boone RB (2020) Enhancing sustainability of grassland ecosystems through ecological restoration and grazing management in an era of climate change on Qinghai-Tibetan Plateau. Agric Ecosyst Environ 287:106684. https://doi.org/10.1016/j.agee.2019.106684

Dotaniya CK, Yashona DS, Aher SB, Rajput PS, Doutaniya RK, Lata M, Mohbe S (2020) Crop performance and soil properties under organic nutrient management. Int J Curr Microbiol App Sci 9(4):1055–1065. https://doi.org/10.20546/ijcmas.2020.904.125

Druilhe Z, Barreiro-Hurlé J (2012) Fertilizer subsidies in sub-Saharan Africa. https://doi.org/10.22004/ag.econ.288997

Edwards CA, Grove TL, Harwood RR, Colfer CP (1993) The role of agroecology and integrated farming systems in agricultural sustainability. Agric Ecosyst Environ 46(1–4):99–121. https://doi.org/10.1016/0167-8809(93)90017-j

El-Shafei AA, Mattar MA (2022) Irrigation scheduling and production of wheat with different water quantities in surface and drip irrigation: field experiments and modelling using CROPWAT and SALTMED. Agronomy 12(7):1488. https://doi.org/10.3390/agronomy12071488

Fageria NK, Baligar VC, Bailey BA (2005) Role of cover crops in improving soil and row crop productivity. Commun Soil Sci Plant Anal 36(19–20):2733–2757. https://doi.org/10.1080/00103620500303939

Fauziyah E, Widyaningsih TS, Kuswantoro DP (2021, December) Management of village treasury land based on bamboo agroforestry. In: IOP conference series: earth and environmental science (Vol 940, No 1, 012007). IOP Publishing. https://doi.org/10.1088/1755-1315/940/1/012007

Fontaine S, Abbadie L, Aubert M, Barot S, Bloor JM, Derrien D et al (2024) Plant–soil synchrony in nutrient cycles: learning from ecosystems to design sustainable agrosystems. Glob Chang Biol 30(1):e17034. https://doi.org/10.1111/gcb.17034

Forister ML, Pelton EM, Black SH (2019) Declines in insect abundance and diversity: we know enough to act now. Conserv Sci Pract 1(8):e80. https://doi.org/10.1111/csp2.80

Francis C, Lieblein G, Gliessman S, Breland TA, Creamer N, Harwood R et al (2003) Agroecology: the ecology of food systems. J Sustain Agric 22(3):99–118. https://doi.org/10.1300/j064v22n03_10

Frelat R, Lopez-Ridaura S, Giller KE, Herrero M, Douxchamps S, Djurfeldt AA et al (2016) Drivers of household food availability in sub-Saharan Africa based on big data from small farms. Proc Natl Acad Sci 113(2):458–463. https://doi.org/10.1073/pnas.1518384112

Gabriel JL, Quemada M, Martín-Lammerding D, Vanclooster M (2019) Assessing the cover crop effect on soil hydraulic properties by inverse modelling in a 10-year field trial. Agric Water Manag 222:62–71. https://doi.org/10.1016/j.agwat.2019.05.034

Gaffney J, Anderson J, Franks C, Collinson S, MacRobert J, Woldemariam W, Albertsen M (2016) Robust seed systems, emerging technologies, and hybrid crops for Africa. Glob Food Sec 9:36–44. https://doi.org/10.1016/j.gfs.2016.06.001

Gebreyess BF, Amare A (2019) Water harvesting technologies in semi-arid and arid areas. J Degraded Mining Lands Manag 7(1):1921. https://doi.org/10.15243/jdmlm.2019.071.1921

Ghasemzadeh A (2012) Global issues of food production. Agrotechnol 1(2):1–2. https://doi.org/10.4172/2168-9881.1000e102

Gill BS, Raupp WJ, Friebe B (2014) Dual threats of imperiled native agroecosystems and climate change to world food security: genomic perspectives. J Crop Improv 28(1):88–98. https://doi.org/10.1080/15427528.2014.865411

Gliessman SR (1995) 3 Sustainable agriculture: an agroecological perspective. In: Advances in plant pathology, vol 11. Academic Press, pp 45–57. https://doi.org/10.1016/s0736-4539(06)80005-x

Gliessman SR (2020) Transforming food and agriculture systems with agroecology. Agric Hum Values 37:547–548. https://doi.org/10.1007/s10460-020-10058-0

Glover JD, Reganold JP, Bell LW, Borevitz J, Brummer EC, Buckler ES et al (2010) Increased food and ecosystem security via perennial grains. Science 328(5986):1638–1639. https://doi.org/10.1126/science.1188761

Grisley W (1993) Seed for bean production in sub-Saharan Africa: issues, problems, and possible solutions. Agric Syst 43(1):19–33. https://doi.org/10.1016/0308-521x(93)90090-o

Hampton JG, Conner AJ, Boelt B, Chastain TG, Rolston P (2016) Climate change: seed production and options for adaptation. Agriculture 6(3):33. https://doi.org/10.3390/agriculture6030033

Havlin J, Heiniger R (2020) Soil fertility management for better crop production. Agronomy 10(9):1349. https://doi.org/10.3390/agronomy10091349

He L, Wang S, Peng C, Tan Q (2018) Optimization of water consumption distribution based on crop suitability in the middle reaches of Heihe River. Sustain For 10(7):2119. https://doi.org/10.3390/su10072119

Hernandez RR, Hoffacker MK, Murphy-Mariscal ML, Wu GC, Allen MF (2015) Solar energy development impacts on land cover change and protected areas. Proc Natl Acad Sci 112(44):13579–13584. https://doi.org/10.1073/pnas.1517656112

Hobbs J, Dozier I, Hovakimyan N (2021) Residue density segmentation for monitoring and optimizing tillage practices. arXiv preprint arXiv:2102.04866. https://www.climatechange.ai/papers/neurips2020/87/paper.pdf

Huang G, Qin S, Zhang S, Cai X, Wu S, Dao J et al (2018) Performance, economics and potential impact of perennial rice PR23 relative to annual rice cultivars at multiple locations in Yunnan Province of China. Sustain For 10(4):1086. https://doi.org/10.3390/su10041086

Hudson N (1987) Soil and water conservation in semi-arid areas (No 57). Food Agric Org https://www.amazon.com/Water-Conservation-Semi-arid-Areas-bulletins/dp/9251026068

Jager HI, Parish ES, Langholtz MH, King AW (2020) Perennials in flood-prone areas of agricultural landscapes: a climate adaptation strategy. Bioscience 70(4):278–280. https://doi.org/10.1093/biosci/biaa006

Jayne TS, Sitko NJ, Mason NM, Skole D (2018) Input subsidy programs and climate smart agriculture: current realities and future potential. Clim Smart Agric 251–273. https://doi.org/10.1007/978-3-319-61194-5_12

Juroszek P, Von Tiedemann A (2011) Potential strategies and future requirements for plant disease management under a changing climate. Plant Pathol 60(1):100–112. https://doi.org/10.1111/j.1365-3059.2010.02410.x

Kanzler M, Böhm C, Freese D (2021) The development of soil organic carbon under young black locust (*Robinia pseudoacacia* L.) trees at a post-mining landscape in eastern Germany. New For 52(1):47–68. https://doi.org/10.1007/s11056-020-09779-1

Katsvairo TW, Wright DL, Marois JJ, Hartzog DL, Rich JR, Wiatrak PJ (2006) Sod–livestock integration into the peanut–cotton rotation: a systems farming approach. Agron J 98(4):1156–1111. https://doi.org/10.2134/agronj2005.0088

Kavian A, Kalehhouei M, Gholami L, Jafarian Z, Mohammadi M, Rodrigo-Comino J (2020) The use of straw mulches to mitigate soil erosion under different antecedent soil moistures. Water 12(9):2518. https://doi.org/10.3390/w12092518

Kindomihou VM (2020) Introductory chapter: milestones in grasses and grassland research. Grasses and Grassland Aspects, 1–30. https://doi.org/10.5772/intechopen.90770

Koocheki A, Gliessman SR (2005) Pastoral nomadism, a sustainable system for grazing land management in arid areas. J Sustain Agric 25(4):113–131. https://doi.org/10.1300/j064v25n04_09

Koskey G, Mburu SW, Awino R, Njeru EM, Maingi JM (2021) Potential use of beneficial microorganisms for soil amelioration, phytopathogen biocontrol, and sustainable crop production in smallholder agroecosystems. Front Sustain Food Syst 5:606308. https://doi.org/10.3389/fsufs.2021.606308

Kumar M, Chaudhary V, Naresh RK, Maurya OP, Pal SL (2018a) Does integrated sources of nutrients enhance growth, yield, quality and soil fertility of vegetable crops. Int J Curr Microb App Sc 7(6):125–155. https://doi.org/10.20546/ijcmas.2018.706.017

Kumar P, Thakur CL, Rai P, Attri K (2018b) Identification of existing agroforestry systems and socio-economic assessment in Kandaghat block of Solan district, Himachal Pradesh, India. Int J Curr Microbiol App Sci 7(4):3815–3826. https://doi.org/10.20546/ijcmas.2018.704.429

Kumar VS, Kumar RP, Madhuri SB, Krishna JNCH (2018c) Effect of microclimate alteration devices and feed additive on reproductive performance of Murrah buffaloes. Pharma Innov J 7(04):2463–2472

Kumawat A, Yadav D, Samadharmam K, Rashmi I (2020) Soil and water conservation measures for agricultural sustainability. Soil Moist Import 23. https://doi.org/10.5772/intechopen.92895

Lacombe C, Couix N, Hazard L (2018) Designing agroecological farming systems with farmers: a review. Agric Syst 165:208–220. https://doi.org/10.1016/j.agsy.2018.06.014

Larsen TA, Hoffmann S, Lüthi C, Truffer B, Maurer M (2016) Emerging solutions to the water challenges of an urbanizing world. Science 352(6288):928–933. https://doi.org/10.1126/science.aad8641

Leakey RR (2017) Trees: a keystone role in agroecosystem function: an update. Multifunctional agriculture–achieving sustainable development in Africa, 19–20. https://doi.org/10.1016/b978-0-12-805356-0.00003-9

Leakey RR (2020) A re-boot of tropical agriculture benefits food production, rural economies, health, social justice and the environment. Nature Food 1(5):260–265. https://doi.org/10.1038/s43016-020-0076-z

Ledo A, Hillier J, Smith P, Aguilera E, Blagodatskiy S, Brearley FQ et al (2019) A global, empirical, harmonised dataset of soil organic carbon changes under perennial crops. Sci Data 6(1):57. https://doi.org/10.1038/s41597-019-0062-1

Lee TG, Lee CG, Hong SG, Kim JH, Park SJ (2019) Water and soil properties in organic and conventional paddies throughout the rice cultivation cycle in South Korea. Environ Eng Res 24(1):45–53. https://doi.org/10.4491/eer.2017.219

Liu YF, Liu Y, Wu GL, Shi ZH (2019) Runoff maintenance and sediment reduction of different grasslands based on simulated rainfall experiments. J Hydrol 572:329–335. https://doi.org/10.1016/j.jhydrol.2019.03.008

Liu L, Zhou W, Guan K, Peng B, Xu S, Tang J et al (2024) Knowledge-guided machine learning can improve carbon cycle quantification in agroecosystems. Nat Commun 15(1):357. https://doi.org/10.1038/s41467-023-43860-5

Lu YC, Watkins KB, Teasdale JR, Abdul-Baki AA (2000) Cover crops in sustainable food production. Food Rev Intl 16(2):121–157. https://doi.org/10.1081/fri-100100285

Maia AG, Miyamoto BCB, Garcia JR (2018) Climate change and agriculture: do environmental preservation and ecosystem services matter? Ecol Econ 152:27–39. https://doi.org/10.1016/j.ecolecon.2018.05.013

Marioti J, Bertol I, Ramos JC, Werner RDS, Padilha J, Bandeira DH (2013) Erosão hídrica em semeadura direta de milho e soja nas direções da pendente e em contorno ao declive, comparada ao solo sem cultivo e descoberto. Rev Bras Ciênc Solo 37:1361–1371. https://doi.org/10.1590/s0100-06832013000500025

Martey E, Kuwornu JK, Adjebeng-Danquah J (2019) Estimating the effect of mineral fertilizer use on Land productivity and income: evidence from Ghana. Land Use Policy 85:463–475. https://doi.org/10.1016/j.landusepol.2019.04.027

Mayer A, Hausfather Z, Jones AD, Silver WL (2018) The potential of agricultural land management to contribute to lower global surface temperatures. Sci Adv 4(8):eaaq0932. https://doi.org/10.1126/sciadv.aaq0932

McGuire S, Sperling L (2016) Seed systems smallholder farmers use. Food Secur 8:179–195. https://doi.org/10.1007/s12571-015-0528-8

Mijangos I, Pérez R, Albizu I, Garbisu C (2006) Effects of fertilization and tillage on soil biological parameters. Enzym Microb Technol 40(1):100–106. https://doi.org/10.1016/j.enzmictec.2005.10.043

Mokany K, Ferrier S, Harwood TD, Ware C, Di Marco M, Grantham HS et al (2020) Reconciling global priorities for conserving biodiversity habitat. Proc Natl Acad Sci 117(18):9906–9911. https://doi.org/10.1073/pnas.1918373117

Morgan JA, Follett RF, Allen LH, Del Grosso S, Derner JD, Dijkstra F et al (2010) Carbon sequestration in agricultural lands of the United States. J Soil Water Conserv 65(1):6A–13A. https://doi.org/10.2489/jswc.65.1.6a

Mus F, Crook MB, Garcia K, Garcia Costas A, Geddes BA, Kouri ED et al (2016) Symbiotic nitrogen fixation and the challenges to its extension to nonlegumes. Appl Environ Microbiol 82(13):3698–3710. https://doi.org/10.1128/aem.01055-16

Nair PR (2007) The coming of age of agroforestry. J Sci Food Agric 87(9):1613–1619. https://doi.org/10.1002/jsfa.2897

Nair PR (2011) Agroforestry systems and environmental quality: introduction. J Environ Qual 40(3):784–790. https://doi.org/10.2134/jeq2011.0076

Nair PKR (2013) Agroforestry: trees in support of sustainable agriculture. In: Sustainable agriculture (pp 270–282). Reference module in earth systems and environmental sciences. https://doi.org/10.1016/b978-0-12-409548-9.05088-0

Nair PKR (2014) Agroforestry: practices and systems, an introduction to agroforestry, 270–282. https://doi.org/10.1016/b978-0-444-52512-3.00021-8

Nayak B, Rath BS, Shahid M, Jena SN, Bagchi TB, Roy PS (2020) Organic nutrient management in aromatic rice-rice sequence: a critical review. Int J Chem Stud 8:1435–1444. https://doi.org/10.22271/chemi.2020.v8.i5t.10503

Niderkorn V, Jayanegara A (2021) Opportunities offered by plant bioactive compounds to improve silage quality, animal health and product quality for sustainable ruminant production: a review. Agronomy 11(1):86. https://doi.org/10.3390/agronomy11010086

van Niekerk J, Wynberg R (2017) Traditional seed and exchange systems cement social relations and provide a safety net: a case study from Kwa Zulu-Natal, South Africa. Agroecol Sustain Food Syst 41(9–10):1099–1123. https://doi.org/10.1080/21683565.2017.1359738

Nirala SK, Suresh R, Kumar S (2018) Fertigation effect on carnation under polyhouse in North Bihar Agro-Climatic Conditions. Int J Curr Microbiol App Sci 7(8):3746–3755. https://doi.org/10.20546/ijcmas.2018.708.380

Nodari RO, Guerra MP (2015) A agroecologia: estratégias de pesquisa e valores. Estudos avançados 29:183–207. https://doi.org/10.1590/s0103-40142015000100010

Nyamangara J, Kodzwa J, Masvaya EN, Soropa G (2020) The role of synthetic fertilizers in enhancing ecosystem services in crop production systems in developing countries. In: The role of ecosystem services in sustainable food systems. Academic Press, pp 95–117. https://doi.org/10.1016/b978-0-12-816436-5.00005-6

de Otálora XD, Epelde L, Arranz J, Garbisu C, Ruiz R, Mandaluniz N (2021) Regenerative rotational grazing management of dairy sheep increases springtime grass production and topsoil carbon storage. Ecol Indic 125:107484. https://doi.org/10.1016/j.ecolind.2021.107484

Outreach S (2007) Managing cover crops profitably. Sustainable agriculture and research education. https://www.amazon.com/Managing-Cover-Crops-Profitably-Outreach/dp/1888626127

Oyana TJ, Kayendeke E, Bamutaze Y, Kisanga D (2015) A field assessment of land use systems and soil properties at varied landscape positions in a fragile ecosystem of Mount Elgon, Uganda. African Geogr Rev 34(1):83–103. https://doi.org/10.1080/19376812.2014.929970

Pagliarino E, Orlando F, Vaglia V, Rolfo S, Bocchi S (2020) Participatory research for sustainable agriculture: the case of the Italian agroecological rice network. Eur J Futures Res 8:1–16. https://doi.org/10.1186/s40309-020-00166-9

Panel MM (2020) Meat, milk and more: policy innovations to shepherd inclusive and sustainable livestock systems in Africa. Intl Food Policy Res Inst. https://doi.org/10.2499/9780896293861

Patanè C, Tringali S, Sortino O (2011) Effects of deficit irrigation on biomass, yield, water productivity and fruit quality of processing tomato under semi-arid Mediterranean climate conditions. Sci Hortic 129(4):590–596. https://doi.org/10.1016/j.scienta.2011.04.030

Patel PP, Champaneri DD (2020) Organic farming: a path to healthy food and environment. Int J Curr Microbiol App Sci 9(03):637–644. https://doi.org/10.20546/ijcmas.2020.903.076

Perakis C, Kyriakarakos G, Hani NB, Hammad S, Damasiotis M (2017) Investigation of solar-powered drip irrigation: the case study of the Jordan Valley-short communication. https://doi.org/10.17221/12/2016-rae

Pinho RCD, Miller RP, Alfaia SS (2012) Agroforestry and the improvement of soil fertility: a view from amazonia. Hindawi Publ Corp Appl Environ Sci 2012:1–11. https://doi.org/10.1155/2012/616383

Pinstrup-Andersen P, Herforth A (2008) Food security: achieving the potential. Environ Sci Policy Sustain Dev 50(5):48–61. https://doi.org/10.3200/envt.50.5.48-61

Power AG (2010) Ecosystem services and agriculture: tradeoffs and synergies. Philos Trans R Soc B Biol Sci 365(1554):2959–2971. https://doi.org/10.1098/rstb.2010.0143

Pretty J, Pervez Bharucha Z (2015) Integrated pest management for sustainable intensification of agriculture in Asia and Africa. Insects 6(1):152–182. https://doi.org/10.3390/insects6010152

Raimi A, Adeleke R, Roopnarain A (2017) Soil fertility challenges and Biofertiliser as a viable alternative for increasing smallholder farmer crop productivity in sub-Saharan Africa. Cogent Food Agric 3(1):1400933. https://doi.org/10.1080/23311932.2017.1400933

Rao RJ (2005) Participatory watershed management (PWM): an approach for integrated development of rural India: a case study from Karnataka, southern India. Int J Environ Technol Manag 5(1):107–115. https://doi.org/10.1504/ijetm.2005.006510

Reddy ARG, Santosh DT, Tiwari KN (2017) Effect of drip irrigation and fertigation on growth, development and yield of vegetables and fruits. Int J Curr Microbiol Appl Sci 6(2):1471–1483. https://doi.org/10.20546/ijcmas.2017.602.165

Reid RS, Fernández-Giménez ME, Galvin KA (2014) Dynamics and resilience of rangelands and pastoral peoples around the globe. Annu Rev Environ Resour 39:217–242. https://doi.org/10.1146/annurev-environ-020713-163329

van Rensburg J (2017, May) "Virtuous cycles" for rural innovation and agri-entrepreneurship development. In: 2017 IST-Africa Week Conference (IST-Africa). IEEE, pp 1–7. https://doi.org/10.23919/istafrica.2017.8102328

Sanam AE, Nur MS, Messakh JJ (2022) Alternative management of river watershed with batulesa method from environmental aspects. Devotion J Res Commun Serv 3(7):633–645. https://doi.org/10.36418/dev.v3i7.156

Sanchez PA, Buresh RJ, Leakey RR (1997) Trees, soils, and food security. Philos Trans R Soc Lond B Biol Sci 352(1356):949–961. https://doi.org/10.1098/rstb.1997.0074

Šarauskis, E., Romaneckas, K., Kumhála, F., & Kriaučiūnienė, Z. (2018). Energy use and carbon emission of conventional and organic sugar beet farming. Journal of Cleaner Production, 201, 428-438. https://doi.org/10.1016/j.jclepro.2018.08.077

Sehgal M, Malik M, Singh RV, Kanojia AK, Singode A (2018) Integrated pest management in rice and its future scope. Int J Curr Microbiol App Sci 7(6):2504–2511. https://doi.org/10.20546/ijcmas.2018.706.297

Senthilkumar M, Ganesh S, Srinivas K, Panneerselvam P, Nagaraja A, Kasinath BL (2017) Fertigation for effective nutrition and higher productivity in banana-a review. Int J Curr Microbiol App Sci 6(7):2104–2122. https://doi.org/10.20546/ijcmas.2017.607.248

Sethuraman G, Mohd Zain NA, Yusoff S, Ng YM, Baisakh N, Cheng A (2021) Revamping ecosystem services through agroecology—the case of cereals. Agriculture 11(3):204. https://doi.org/10.3390/agriculture11030204

Shadeed S, Judeh T, Riksen M (2020) Rainwater harvesting for sustainable agriculture in high water-poor areas in the West Bank. Palestine Water 12(2):380. https://doi.org/10.3390/w12020380

Siegner A, Sowerwine J, Acey C (2018) Does urban agriculture improve food security? Examining the nexus of food access and distribution of urban produced foods in the United States: a systematic review. Sustain For 10(9):2988. https://doi.org/10.3390/su10092988

SourceWatch (2008) Association of college and research libraries, 45(11), 45–5895. Choice Reviews Online https://doi.org/10.5860/choice.45-5895

Sperling L, Gallagher P, McGuire S, March J, Templer N (2020) Informal seed traders: the backbone of seed business and African smallholder seed supply. Sustain For 12(17):7074. https://doi.org/10.3390/su12177074

Srivastav AL (2020) Chemical fertilizers and pesticides: role in groundwater contamination. In: Agrochemicals detection, treatment and remediation. Butterworth-Heinemann, pp 143–159. https://doi.org/10.1016/b978-0-08-103017-2.00006-4

Stroosnijder L, Moore D, Alharbi A, Argaman E, Biazin B, van den Elsen E (2012) Improving water use efficiency in drylands. Curr Opin Environ Sustain 4(5):497–506. https://doi.org/10.1016/j.cosust.2012.08.011

Suárez-Torres J, Suárez-López JR, López-Paredes D, Morocho H, Cachiguango-Cachiguango LE, Dellai W (2017) Agroecology and health: Lessons from indigenous populations. Curr Environ Health Reports 4:244–251. https://doi.org/10.1007/s40572-017-0146-z

Swami S (2020) Soil health management under organic production system. IJCS 8(2):330–339. https://doi.org/10.22271/chemi.2020.v8.i2e.8789

Swift RS (2001) Sequestration of carbon by soil. Soil Sci 166(11):858–871. https://doi.org/10.1097/00010694-200111000-00010

Tadesse B, Beyene F, Kassa W, Wentzell R (2015) The roles of customary institutions in adaptation and coping to climate change and variability among the Issa, Ittu and Afar Pastoralists of eastern Ethiopia. Change Adapt Socio-ecol Syst 2(1). https://doi.org/10.1515/cass-2015-0025

Tamburini G, De Simone S, Sigura M, Boscutti F, Marini L (2016) Soil management shapes ecosystem service provision and trade-offs in agricultural landscapes. Proc R Soc B Biol Sci 283(1837):20161369. https://doi.org/10.1098/rspb.2016.1369

Teague WR (2018) Forages and pastures symposium: Cover crops in livestock production: whole-system approach: managing grazing to restore soil health and farm livelihoods. J Anim Sci 96(4):1519–1530. https://doi.org/10.1093/jas/skx060

Tenywa MM, Bekunda M (2009) Managing soils in Sub-Saharan Africa: challenges and opportunities for soil and water conservation. J Soil Water Conserv 64(1):44A–48A. https://doi.org/10.2489/jswc.64.1.44a

Thrupp LA (2000) Linking agricultural biodiversity and food security: the valuable role of agrobiodiversity for sustainable agriculture. Int Aff 76(2):265–281. https://doi.org/10.1111/1468-2346.00133

Tung OJL (2018) African organic product standards for the African continent? Prospects and limitations. Potchefstroom Electronic Law Journal/Potchefstroomse Elektroniese Regsblad 21(1). https://doi.org/10.17159/1727-3781/2018/v21i0a4308

Venkateshwaran M, Ané JM (2011) Legumes and nitrogen fixation: physiological, molecular, evolutionary perspective and applications. In: The molecular basis of nutrient use efficiency in crops, pp 457–489. https://doi.org/10.1002/9780470960707.ch20

Verchot LV, Van Noordwijk M, Kandji S, Tomich T, Ong C, Albrecht A et al (2007) Climate change: linking adaptation and mitigation through agroforestry. Mitig Adapt Strateg Glob Chang 12:901–918. https://doi.org/10.1007/s11027-007-9105-6

Verma S, Singh A, Pradhan SS, Singh RK, Singh JP (2017) Bio-efficacy of organic formulations on crop production – a review. Int J Curr Microbiol App Sci 6(5):648–665. https://doi.org/10.20546/ijcmas.2017.605.075

Volkohon VV, Dimova SB, Volkohon KI, Horban VP, Zhurba MA, Shtanko NP et al (2016) Biological nitrogen transformation in agrocenoses of potato and productivity of culture in organic agriculture. Interdepartmental thematic scientific collection. Agric Microbiol 24:3–8. https://doi.org/10.35868/1997-3004.24.3-8

Wade MR, Gurr GM, Wratten SD (2008) Ecological restoration of farmland: progress and prospects. Philos Trans R Soc B Biol Sci 363(1492):831–847. https://doi.org/10.1098/rstb.2007.2186

Weißhuhn P, Reckling M, Stachow U, Wiggering H (2017) Supporting agricultural ecosystem services through the integration of perennial polycultures into crop rotations. Sustain For 9(12):2267. https://doi.org/10.3390/su9122267

Wilson MH, Lovell ST (2016) Agroforestry—the next step in sustainable and resilient agriculture. Sustain For 8(6):574. https://doi.org/10.3390/su8060574

Wisner B (2010) Climate change and cultural diversity. Int Soc Sci J 61(199):131–140. https://doi.org/10.1111/j.1468-2451.2010.01752.x

Xiong W, Tarnavsky E (2020) Better agronomic management increases climate resilience of maize to drought in Tanzania. Atmos 11(9):982. https://doi.org/10.3390/atmos11090982

Yang L, Chen C (2021, March) Research progress on driving factors of species diversity of grassland insects. In: IOP Conference series: earth and environmental science (Vol 692, No 3, p 032094). IOP Publishing. https://doi.org/10.1088/1755-1315/692/3/032094

Zheng B, Campbell JB, Serbin G, Galbraith JM (2014) Remote sensing of crop residue and tillage practices: present capabilities and future prospects. Soil Tillage Res 138:26–34. https://doi.org/10.1016/j.still.2013.12.009

Zomer RJ, Bossio DA, Sommer R, Verchot LV (2017) Global sequestration potential of increased organic carbon in cropland soils. Sci Rep 7(1):15554. https://doi.org/10.1038/s41598-017-15794-8

Chapter 9
Biodiversity Conservation

Sub-Saharan Africa is renowned for its exceptional biodiversity, encompassing diverse ecosystems, including tropical rainforests, savannahs, wetlands, and coastal areas, which support an immense variety of plant and animal species (Norris et al. 2010). Within this context, integrating biodiversity conservation principles into agricultural practices is crucial to ensure sustainable food production and environmental stewardship.

Agroecology recognizes the interconnectedness between agricultural production and biodiversity conservation (Altieri 1999). It acknowledges that agricultural activities have the potential to either contribute to the degradation of natural habitats or support their preservation and restoration (Thrupp 2000). By adopting agroecological practices, farmers in sub-Saharan Africa can actively contribute to the conservation of biodiversity within and around their farmland (Batáry et al. 2011).

Agroecology promotes the implementation of biodiversity-friendly practices, such as maintaining natural vegetation buffers, establishing agroforestry systems, and preserving or restoring riparian zones. These practices help protect and enhance habitats for a wide range of plant and animal species, including pollinators, beneficial insects, birds, and small mammals (Cozim-Melges et al. 2024). The preservation of diverse habitats within agricultural landscapes supports the establishment of ecological corridors and promotes the movement and genetic exchange of species, contributing to overall ecosystem health and resilience.

Agricultural landscapes in sub-Saharan Africa are often intertwined with natural habitats, making it necessary to consider the conservation of biodiversity in farming systems. Agroecologists have emphasized the importance of managing agroecosystems to enhance biodiversity across different biomes, with a range of practices that can contribute to this goal (Cozim-Melges et al. 2024). Strategies such as integrating ecological pest and soil management, establishing agroforestry systems, and protecting natural vegetation buffers can play a crucial role in preserving and enhancing biodiversity within agricultural landscapes (Thrupp 2000).

M. Otieno, *Sustainable Agroecological Practices in Sub-Saharan Africa in the Face of Climate Change*, Advances in Geographical and Environmental Sciences, https://doi.org/10.1007/978-3-031-70472-7_9

9.1 The Role of Habitat Preservation and Restoration in Agroecological Systems

Agroecological approaches recognize the critical importance of preserving and restoring natural habitats within and around agricultural landscapes (Wade et al. 2008). These habitats, such as wetlands, forests, and riparian zones, not only serve as homes for diverse plant and animal species but also provide essential ecological services that support sustainable agricultural production (Wade et al. 2008; Bhagwat et al. 2008; Cozim-Melges et al. 2024).

The preservation and restoration of these natural habitats are vital for the success of agroecology. Wetlands, for instance, help regulate water flow, enhance water quality, and provide habitat for aquatic organisms (Gliessman 1995; Batáry et al. 2011). Forests, on the other hand, offer crucial services such as carbon sequestration, soil erosion control, and the maintenance of microclimate conditions. Similarly, riparian zones contribute to water filtration, flood regulation, and the preservation of aquatic biodiversity (Teague 2018).

Maintaining and restoring these natural habitats within and around agricultural landscapes has several benefits. Foremost, it fosters the conservation of plant and animal species, contributing to the maintenance of biodiversity (Batáry et al. 2011). Biodiversity is crucial for sustainable agriculture as it supports pollination, pest control, and nutrient cycling processes, ultimately enhancing crop productivity and resilience. Additionally, the protection of natural habitats helps preserve the ecosystem services that are essential for the long-term viability of agroecosystems (Gliessman 1995; Batáry et al. 2011; Cozim-Melges et al. 2024).

Accordingly, agroecological approaches prioritize the integration of natural habitats within agricultural landscapes, recognizing their crucial role in promoting a balanced and sustainable agroecosystem. This approach not only supports biodiversity conservation but also ensures the long-term provision of the essential ecosystem services that underpin the productivity and resilience of agroecological systems.

9.2 The Importance of Crop Diversity in Agroecological Systems

Agroecology, as an approach to sustainable agriculture, emphasizes the promotion of diverse crop species and varieties, including traditional and heirloom varieties, to enhance biodiversity and create resilient agricultural systems. Cultivating a variety of crops can support beneficial insects, birds, and other wildlife by providing alternative food sources and habitats within the agricultural landscape (Jat et al. 2021). This diversity plays a crucial role in reducing the vulnerability of agricultural systems to pest and disease outbreaks, as monoculture systems are more susceptible to rapid buildup and spread of pests and pathogens (Altieri 1999; Erisman et al. 2016; Jat et al. 2021).

The utilization of diverse crop species contributes to the establishment of resilient and sustainable agricultural systems. Research has shown that the resilience of agricultural systems increases with a larger biodiversity (Erisman et al. 2016), as diverse crops provide less concentrated and less predictable food sources for pests, making it more difficult for them to establish and spread (Forister et al. 2019). Traditional farming systems are often repositories of a wealth of principles and measures that can help modern agricultural systems become more resilient to climatic extremes, such as crop diversification, maintaining local genetic diversity, and water conservation (Altieri et al. 2015).

Moreover, the genetic diversity inherent in traditional and heirloom crop varieties often possess traits that are well-adapted to local conditions and can attract and support a wide range of beneficial insects, including pollinators and natural enemies of pests (Altieri 1999; Altieri et al. 2015; Erisman et al. 2016). This diversity within agroecosystems performs a variety of ecological services beyond food production, such as recycling of nutrients, regulation of microclimate, and pest and disease regulation (Altieri 1999; Altieri et al. 2015; Erisman et al. 2016). Implementing agroecological practices that enhance biodiversity, such as polycultures, agroforestry, and crop-livestock mixed systems, can strengthen the resilience of farmers and rural communities in the face of climate change (Altieri et al. 2015; Forister et al. 2019).

9.3 The Role of Agroforestry in Promoting Biodiversity and Ecosystem Services

Agroforestry systems have emerged as a promising approach to reconciling agricultural production with environmental conservation (Bhagwat et al. 2008). These integrated systems involve the intentional combination of trees and crops, creating diverse and multi-layered landscapes that can provide a wide range of benefits (Bhagwat et al. 2008; Nair 2007).

Agroforestry practices, such as alley cropping, windbreaks, and shade-grown agriculture, play a crucial role in enhancing biodiversity, improving soil fertility, and delivering essential ecosystem services (Nair 2007; Bhagwat et al. 2008). Alley cropping, for instance, involves cultivating crops between rows of trees, which can reduce evapotranspiration, conserve soil moisture, and enhance soil structure and nutrient cycling, leading to improved ecosystem health (Nair 2007; Kanzler et al. 2021; Fauziyah et al. 2021). The presence of trees in these systems also provides habitat and food sources for a variety of plant and animal species, increasing overall biodiversity. Similarly, windbreaks, which are linear plantings of trees or shrubs, can protect crops from wind erosion, minimize evapotranspiration, and provide shelter for livestock, while also contributing to biodiversity conservation by offering habitats for birds, insects, and small mammals (Nair 2007) Shade-grown agriculture, another agroforestry practice, can also support biodiversity by providing suitable environments for a range of species (Nair 2007; Kanzler et al. 2021; Fauziyah et al. 2021). Across these systems, the integration of trees and agricultural activities can enhance carbon sequestration, water regulation, and other ecosystem services, offering a more sustainable alternative to monoculture production (Nair 2007; Kanzler et al. 2021; Fauziyah et al. 2021). While more research and effort are needed to fully explore the potential of agroforestry, the existing evidence suggests that these integrated systems can play a crucial role in promoting biodiversity and delivering a wide range of environmental benefits (Nair 2007; Kanzler et al. 2021).

9.4 The Importance of Conserving Indigenous and Traditional Crop Varieties in Agroecology

Agroecology places significant emphasis on the conservation and utilization of indigenous and traditional crop varieties that have coevolved with local ecosystems (Thrupp 2000; Shen et al. 2017). These crop varieties, developed through generations of farmers' selection and adaptation, possess a wealth of genetic diversity and adaptive traits that are essential for ensuring long-term food security and resilience in the face of changing climatic conditions. By preserving and promoting the cultivation of these indigenous and traditional crop varieties, agroecology contributes to the conservation of agricultural biodiversity.

Indigenous and traditional crop varieties are often well-suited to the local agroecological conditions, as they have been selected and adapted to thrive in specific ecological niches. Their genetic diversity allows them to possess a wide range of traits, such as resistance to pests and diseases, tolerance to drought, and adaptation to different soil types, which are crucial for withstanding environmental challenges and maintaining agricultural productivity in the face of climatic uncertainties (Thrupp 2000; Gill et al. 2014; Altieri et al. 2015; Flint-Garcia et al. 2023). In contrast, modern high-yielding varieties, which dominate industrial agriculture, often lack the genetic diversity and adaptability of indigenous and traditional varieties, making agricultural systems more vulnerable to pests, diseases, and changing environmental conditions.

Conserving and promoting indigenous and traditional crop varieties aligns with the principles of agroecology, which aim to enhance the resilience and sustainability of farming systems by diversifying crop species and varieties. The utilization of diverse crop varieties contributes to the maintenance of a diverse agroecosystem, which is a key strategy for adapting to the impacts of climate change (Gill et al. 2014; Altieri et al. 2015). Recent research has shown that increasing crop rotational diversity can enhance cereal yields, demonstrating the benefits of diversifying agricultural production systems (Smith et al. 2023). Furthermore, understanding the agroecological features that underlie the resilience of traditional agroecosystems is crucial, as they can serve as the foundation for the design of climate change-resilient farming systems of the future (Gupta 2010; Gill et al. 2014; Altieri et al. 2015).

Indigenous variety of pigeon pea (*Cajanus cajan*)

9.5 Promoting Pollinators and Beneficial Insects in Agroecological Systems

Agroecology recognizes the critical role of pollinators and beneficial insects in maintaining the sustainability and resilience of agricultural systems (Potts et al. 2010; Isaacs et al. 2017; Deutsch et al. 2018). Pollinators, such as bees, butterflies, and other insects, are essential for the transfer of pollen between flowers, facilitating the fertilization and reproduction of many crop plants (Potts et al. 2010; Thapa 2006). Beneficial insects, including predators and parasitoids, help control pest populations by preying on or parasitizing them, reducing the need for chemical pesticides (Deutsch et al. 2018). By prioritizing the conservation and support of these important insect groups, agroecological practices contribute to the overall ecosystem functioning (Kearns et al. 1998).

A nesting site for bees and other beneficial arthropods within the farm

Agroecology promotes the creation of suitable habitats and provision of nesting sites for pollinators and beneficial insects within agricultural landscapes. This can be achieved through various practices, such as planting diverse flowering plants, establishing hedgerows or field margins, and incorporating insect-friendly elements like insectary plants or nesting structures. These habitat enhancements provide valuable food resources, shelter, and nesting opportunities, supporting the populations of pollinators and beneficial insects and enhancing their ecosystem services in agricultural areas (Deutsch et al. 2018; Forister et al. 2019).

Furthermore, agroecology emphasizes minimizing the use of broad-spectrum chemical pesticides that can negatively impact both target and non-target insect populations. By adopting integrated pest management (IPM) approaches that combine cultural, biological, and ecological pest control strategies, agroecological systems reduce the reliance on chemical pesticides and allow insect populations to thrive. These practices contribute to the sustainability and resilience of agricultural systems by maintaining the crucial roles of pollinators and beneficial insects.

9.6 Integrating Participatory Approaches and Indigenous Knowledge in Agroecology

Agroecology is a holistic approach to sustainable agriculture that prioritizes the active involvement of farmers, local communities, and indigenous peoples in decision-making processes related to agricultural and land management practices (Altieri 2002). This inclusive approach acknowledges the wealth of knowledge and experience held by these groups and recognizes their role as custodians of biodiversity. By integrating traditional knowledge and practices, agroecology not only benefits from the accumulated wisdom of generations but also contributes to the conservation of traditional agroecosystems and the biodiversity they support (Suárez-Torres et al. 2017).

Indigenous communities have developed intricate relationships with their local environments over centuries, resulting in the establishment of diverse and resilient traditional agroecosystems (Thrupp 2000). These agroecosystems, characterized by a mosaic of cultivated fields, fallows, forests, and other natural or semi-natural areas, provide habitats for a wide range of plant and animal species (Altieri 2002). They also exhibit unique practices and techniques adapted to local conditions, such as traditional seed saving, agroforestry, and water management systems. Agroecology recognizes the value of these traditional agroecosystems as repositories of biodiversity and as models of sustainable land use.

By actively involving indigenous communities in decision-making processes, agroecology acknowledges their traditional knowledge systems and ensures the continuation and preservation of these practices. Indigenous knowledge is based on a deep understanding of local ecosystems and the relationships between different elements, such as crops, livestock, wildlife, and natural resources. This knowledge often includes sophisticated observation and prediction of weather patterns, pest and disease cycles, and the selection and management of diverse crop varieties and animal breeds (Thrupp 2000; Altieri 2002; Leakey 2017; Guiriba 2019).

Agroecological and participatory approaches have been shown to be particularly useful for agrarian systems, as they provide information about their physical functioning and their spatial/temporal differences, highlighting the diversity that exists in the structure and physical/biological functioning of agriculture (Asrat 2018).

This approach can help guide coherent decision-making regarding resource-use efficiency in rural development processes.

By embracing participatory approaches and integrating indigenous knowledge, agroecology can foster more resilient and biodiverse agroecosystems that provide a wide range of ecosystem services and benefits to local communities (Thrupp 2000; Cozim-Melges et al. 2024). Agroecology has been recognized as a stepping stone to achieving several Sustainable Development Goals, due to its great potential to build climate change-resilient farming systems while enhancing ecosystem services and reducing biodiversity loss (Gliessman 1995; Asrat 2018; Gliessman 2020; Sethuraman et al. 2021).

References

Altieri MA (1999) The ecological role of biodiversity in agroecosystems. Agric Ecosyst Environ 74(1–3):19–31. https://doi.org/10.1016/s0167-8809(99)00028-6

Altieri MA (2002) Agroecology: the science of natural resource management for poor farmers in marginal environments. Agric Ecosyst Environ 93(1–3):1–24. https://doi.org/10.1016/s0167-8809(02)00085-3

Altieri MA, Nicholls CI, Henao A, Lana M (2015) Agroecology and the design of climate change-resilient farming systems. Agron Sustain Dev 35(3):869–890. https://doi.org/10.1007/s13593-015-0285-2

Asrat GA (2018) Mapping soil terrain resources and descriptions of agro-ecological zone in Dawuro and Gamo Gofa zones in south-western Ethiopia. Soil Sci Environ Manag J 9(10):164–179. https://doi.org/10.5897/jssem2018.0695

Batáry P, Báldi A, Kleijn D, Tscharntke T (2011) Landscape-moderated biodiversity effects of agri-environmental management: a meta-analysis. Proc R Soc B Biol Sci 278(1713):1894–1902. https://doi.org/10.1098/rspb.2010.1923

Bhagwat SA, Willis KJ, Birks HJB, Whittaker RJ (2008) Agroforestry: a refuge for tropical biodiversity? Trends Ecol Evol 23(5):261–267. https://doi.org/10.1016/j.tree.2008.01.005

Cozim-Melges F, Ripoll-Bosch R, Veen GF, Oggiano P, Bianchi FJ, van der Putten WH, van Zanten HH (2024) Farming practices to enhance biodiversity across biomes: a systematic review. NPJ Biodivers 3(1):1. https://doi.org/10.1038/s44185-023-00034-2

Deutsch CA, Tewksbury JJ, Tigchelaar M, Battisti DS, Merrill SC, Huey RB, Naylor RL (2018) Increase in crop losses to insect pests in a warming climate. Science 361(6405):916–919. https://doi.org/10.1126/science.aat3466

Erisman JW, van Eekeren N, de Wit J, Koopmans C, Cuijpers W, Oerlemans N, Koks BJ (2016) Agriculture and biodiversity: a better balance benefits both. AIMS Agric Food 1(2):157–174. https://doi.org/10.3934/agrfood.2016.2.157

Fauziyah E, Widyaningsih TS, Kuswantoro DP (2021, December) Management of village treasury land based on bamboo agroforestry. In: IOP conference series: earth and environmental science (Vol 940, No 1, 012007). IOP Publishing. https://doi.org/10.1088/1755-1315/940/1/012007

Flint-Garcia S, Feldmann MJ, Dempewolf H, Morrell PL, Ross-Ibarra J (2023) Diamonds in the not-so-rough: wild relative diversity hidden in crop genomes. PLoS Biol 21(7):e3002235. https://doi.org/10.1371/journal.pbio.3002235

Forister ML, Pelton EM, Black SH (2019) Declines in insect abundance and diversity: we know enough to act now. Conserv Sci Pract 1(8):e80. https://doi.org/10.1111/csp2.80

Gill BS, Raupp WJ, Friebe B (2014) Dual threats of imperiled native agroecosystems and climate change to world food security: genomic perspectives. J Crop Improv 28(1):88–98. https://doi.org/10.1080/15427528.2014.865411

Gliessman SR (1995) 3 Sustainable agriculture: an agroecological perspective. In: Advances in plant pathology, vol 11. Academic Press, pp 45–57. https://doi.org/10.1016/s0736-4539(06)80005-x
Gliessman SR (2020) Transforming food and agriculture systems with agroecology. Agric Hum Values 37:547–548. https://doi.org/10.1007/s10460-020-10058-0
Guiriba GO (2019) Documentation of indigenous knowledge on production and post-harvest management of sweet potato in the bicol region. Philippines Hasanuddin Univ 3(1):93–93. https://doi.org/10.20956/jars.v3i1.1719
Gupta J (2010) A history of international climate change policy. Wiley Interdiscip Rev Clim Chang 1(5):636–653. https://doi.org/10.1002/wcc.67
Isaacs R, Williams N, Ellis J, Pitts-Singer TL, Bommarco R, Vaughan M (2017) Integrated crop pollination: combining strategies to ensure stable and sustainable yields of pollination-dependent crops. Basic Appl Ecol 22:44–60. https://doi.org/10.1016/j.baae.2017.07.003
Jat SL, Suby SB, Parihar CM, Gambhir G, Kumar N, Rakshit S (2021) Microbiome for sustainable agriculture: a review with special reference to the corn production system. Arch Microbiol 203(6):2771–2793. https://doi.org/10.1007/s00203-021-02320-8
Kanzler M, Böhm C, Freese D (2021) The development of soil organic carbon under young black locust (*Robinia pseudoacacia* L.) trees at a post-mining landscape in eastern Germany. New For 52(1):47–68. https://doi.org/10.1007/s11056-020-09779-1
Kearns CA, Inouye DW, Waser NM (1998) Endangered mutualisms: the conservation of plant-pollinator interactions. Annu Rev Ecol Syst 29(1):83–112. https://doi.org/10.1146/annurev.ecolsys.29.1.83
Leakey RR (2017) Trees: a keystone role in agroecosystem function: an update. Multifunctional agriculture–achieving sustainable development in Africa, 19–20. https://doi.org/10.1016/b978-0-12-805356-0.00003-9
Nair PR (2007) The coming of age of agroforestry. J Sci Food Agric 87(9):1613–1619. https://doi.org/10.1002/jsfa.2897
Norris K, Asase A, Collen B, Gockowksi J, Mason J, Phalan B, Wade A (2010) Biodiversity in a forest-agriculture mosaic – the changing face of West African rainforests. Biol Conserv 143(10):2341–2350. https://doi.org/10.1016/j.biocon.2009.12.032
Potts SG, Biesmeijer JC, Kremen C, Neumann P, Schweiger O, Kunin WE (2010) Global pollinator declines: trends, impacts and drivers. Trends Ecol Evol 25(6):345–353. https://doi.org/10.1016/j.tree.2010.01.007
Sethuraman G, Mohd Zain NA, Yusoff S, Ng YM, Baisakh N, Cheng A (2021) Revamping ecosystem services through agroecology—the case of cereals. Agriculture 11(3):204. https://doi.org/10.3390/agriculture11030204
Shen S, Xu G, Li D, Clements DR, Zhang F, Jin G et al (2017) Agrobiodiversity and in situ conservation in ethnic minority communities of Xishuangbanna in Yunnan Province, Southwest China. J Ethnobiol Ethnomed 13:1–15. https://doi.org/10.1186/s13002-017-0158-7
Smith ME, Vico G, Costa A, Bowles T, Gaudin AC, Hallin S et al (2023) Increasing crop rotational diversity can enhance cereal yields. Commun Earth Environ 4(1):89. https://doi.org/10.1038/s43247-023-00746-0
Suárez-Torres J, Suárez-López JR, López-Paredes D, Morocho H, Cachiguango-Cachiguango LE, Dellai W (2017) Agroecology and health: Lessons from indigenous populations. Curr Environ Health Reports 4:244–251. https://doi.org/10.1007/s40572-017-0146-z
Teague WR (2018) Forages and pastures symposium: Cover crops in livestock production: whole-system approach: managing grazing to restore soil health and farm livelihoods. J Anim Sci 96(4):1519–1530. https://doi.org/10.1093/jas/skx060
Thapa RB (2006) Honeybees and other insect pollinators of cultivated plants: a review. J Inst Agric Anim Sci 27:1–23. https://doi.org/10.3126/jiaas.v27i0.691

Thrupp LA (2000) Linking agricultural biodiversity and food security: the valuable role of agrobiodiversity for sustainable agriculture. Int Aff 76(2):265–281. https://doi.org/10.1111/1468-2346.00133

Wade MR, Gurr GM, Wratten SD (2008) Ecological restoration of farmland: progress and prospects. Philos Trans R Soc B Biol Sci 363(1492):831–847. https://doi.org/10.1098/rstb.2007.2186

Chapter 10
Farm Health: Crop Health, Animal Health, and Food Quality

The maintenance of farm health, encompassing the well-being of crops, animals, and the quality of produced food, is a vital aspect of sustainable agriculture. Sustainable food production not only optimizes health, safety, and quality but also minimizes the environmental impact through reduced inputs of energy, fertilizers, pesticides, and water, as well as minimizing waste generation and emissions (Seiber and Kleinschmidt 2014).

Soil and crop management strategies are crucial in ensuring higher crop productivity within a sustainable environment (Shah and Wu 2019). The importance of soil health management practices cannot be overstated, as they directly impact the prevalence of soil-borne pathogens, nematodes, and root diseases, which can significantly affect the production of vegetables and other food crops (Abawi and Widmer 2000; Seiber and Kleinschmidt 2014; Shah and Wu 2019; Selladurai et al. 2020). Developing a comprehensive understanding of the impact of specific farming practices on soil properties and quality under various climatic conditions is essential for the long-term sustainability of agricultural systems. Furthermore, the integration of scientific knowledge, economic considerations, and social costs of such practices that enhance soil quality must be better understood and incorporated into the development of widely implementable practices.

Crop productivity and ecosystem health are strongly influenced by soil physical quality, particularly in the root zone. The soil must provide a suitable foundation that supports plant growth while also allowing for the flourishing of soil flora and fauna, minimizing nutrient leaching and pesticide runoff, and enabling timely agricultural activities without damaging soil structure or compaction.

Conventional agricultural production systems, which are often chemical and tillage-intensive, tend to degrade soil health due to their negative impact on the biological, chemical, and physical properties of the soil. In contrast, good management practices, such as crop rotations, cover crops, and reduced tillage, can yield

M. Otieno, *Sustainable Agroecological Practices in Sub-Saharan Africa in the Face of Climate Change*, Advances in Geographical and Environmental Sciences, https://doi.org/10.1007/978-3-031-70472-7_10

positive economic and environmental benefits to soil ecosystems, promoting high sustainability and productivity along with healthier soils.

Such practices not only improve soil quality but also contribute to the overall health and productivity of the farm system, including animal health and food quality. The abundance and diversity of beneficial soil microbes, as well as the suppression of crop pests, are essential indicators of soil health and can directly impact the incidence and severity of root diseases, which threaten the long-term sustainability of the agricultural system (Topp et al. 1997; Abawi and Widmer 2000; Arriaga et al. 2017; Selladurai et al. 2020).

10.1 Enhancing Farm Health Through Agroecological Practices in Sub-Saharan Africa

Agroecology is a crucial approach for promoting farm health and establishing sustainable farming systems in sub-Saharan Africa (Leakey 2020). Agroecological practices aim to integrate ecological principles and agricultural techniques to optimize productivity while minimizing negative environmental impacts (Asrat 2018; Leakey 2020; Sethuraman et al. 2021). One of the key objectives of agroecology is to enhance farm health by adopting practices that support ecological processes and reduce the reliance on external inputs such as synthetic pesticides and fertilizers.

Maintaining and improving soil health is essential for sustainable agriculture in the region. Healthy soils contribute to enhanced nutrient cycling, improved moisture retention, and increased resistance to erosion, all of which lead to improved crop productivity (Cozim-Melges et al. 2024). Agroecological approaches, including cover cropping, crop rotation, and agroforestry, are known to enhance soil fertility, structure, and biological activity (Fontaine et al. 2024). These practices can help build ecological resilience in the face of climate change and other external stressors.

Beyond soil health, agroecological practices promote diversified cropping systems and the use of locally adapted crop varieties. This approach enhances biodiversity within the farming system and reduces the vulnerability of crops to pests and diseases. Integrating livestock into the farming system through techniques like rotational grazing can also contribute to farm health by improving nutrient cycling, weed control, and pest management (Leakey 2020; Sethuraman et al. 2021; Fontaine et al. 2024; Cozim-Melges et al. 2024).

Agroecology recognizes the importance of ecological resilience in the face of climate change and other external stressors. By promoting biodiversity, crop diversification, and the use of agroecological principles, farmers can build resilient farming systems that are better equipped to withstand and recover from environmental challenges (Sethuraman et al. 2021; Fontaine et al. 2024; Cozim-Melges et al. 2024). Exploiting the unique environment-tolerant properties of microorganisms, their genetic diversity, and their interactions with plants could also be crucial in addressing the management of abiotic stress in agriculture (Koskey et al. 2021).

10.2 Agroecological Approaches to Improving Crop Health in Sub-Saharan Africa

Crop health and agroecology are pivotal in promoting sustainable agriculture and ensuring food security in sub-Saharan Africa. This region encounters a range of agricultural challenges, such as soil degradation, pests and diseases, climate change impacts, and limited resource accessibility. By implementing agroecological practices and prioritizing crop health, these challenges can be addressed, and agricultural productivity can be improved in a sustainable and environmentally friendly manner. Sub-Saharan Africa faces significant soil degradation, which negatively impacts crop productivity and agricultural sustainability (Mapfumo et al. 2017; Devenish et al. 2023). Soil erosion, nutrient depletion, and soil compaction are prevalent issues in the region (Devenish et al. 2023). Agroecology emphasizes soil conservation and restoration through practices such as conservation agriculture, organic matter incorporation, and agroforestry, which promote soil health, enhance soil structure, improve water retention capacity, and mitigate erosion, thereby contributing to sustainable crop production (Tenywa and Bekunda 2009).

Pests and diseases pose a significant threat to crop yields in sub-Saharan Africa, as the region's diverse agroecosystems provide a favorable environment for their proliferation. Agroecological approaches focus on pest and disease management strategies that rely on ecological processes and minimize reliance on synthetic pesticides, promoting biodiversity, crop diversification, and the use of resistant crop varieties to reduce pest and disease pressure and improve crop health and yield stability. Climate change exacerbates agricultural challenges in sub-Saharan Africa, leading to increased temperature, erratic rainfall patterns, and extreme weather events. Agroecological practices contribute to climate resilience by enhancing the adaptive capacity of farming systems, such as through the use of drought-tolerant crop varieties, soil moisture conservation techniques, and diversified cropping systems that can better withstand these climatic stresses (Jayne et al. 2018; Jackson et al. 2018; Gugissa et al. 2022; Devenish et al. 2023;). Implementing agroecological approaches to address the multifaceted challenges in sub-Saharan Africa's agricultural sector can promote crop health, improve yields, and contribute to the region's overall food security and sustainable development (Tenywa and Bekunda 2009; Jayne et al. 2018; Gugissa et al. 2022; Devenish et al. 2023).

10.3 Promoting Animal Health and Agroecology in Sub-Saharan Africa

Sustainable and resilient agricultural systems in sub-Saharan Africa can be significantly enhanced by promoting animal health and integrating agroecology principles. Animal health is a crucial component of agricultural productivity, food security, and rural livelihoods in the region. By adopting agroecological approaches that prioritize the well-being of livestock and the ecological balance of farming systems, sub-Saharan Africa can achieve long-term sustainability in its agriculture sector.

Livestock plays a vital role in the agricultural economy of sub-Saharan Africa, providing a source of income, food, and draught power for millions of smallholder farmers. However, the region faces various challenges related to animal health, including diseases, inadequate nutrition, and limited access to veterinary services. These challenges contribute to reduced productivity, increased mortality rates, and economic losses for farmers (Panel 2020). Promoting animal health through agro-ecological approaches can help address these challenges and improve livestock productivity sustainably (Alemu et al. 2019; MacCarthy et al. 2021).

Agroecology emphasizes the integration of livestock within farming systems, focusing on practices that enhance ecological processes and minimize external inputs. By integrating livestock, such as cattle, goats, or poultry, farmers can improve nutrient cycling, weed control, and pest management within their agricultural landscapes (Dominguez-Salas et al. 2019; Leakey 2020). Practices like rotational grazing, where animals are moved between different pastures, can promote the health of grazing lands and minimize overgrazing, leading to improved forage quality and quantity. Furthermore, agroecological approaches prioritize diversified farming systems, including mixed-crop livestock production. Integrating crops and livestock on the same farm allows for the recycling of nutrients and the optimization of resource use (Chianu and Tsujii 2004; Panel 2020; Niderkorn and Jayanegara 2021). Animal manure can serve as an organic fertilizer, improving soil fertility and reducing the need for synthetic fertilizers (Chianu and Tsujii 2004; Leakey 2020).

A healthy cow feeding on fresh forage

10.4 The Impact of Agroecological Practices on Food Quality in Sub-Saharan Africa

The integration of agroecology principles in sub-Saharan Africa holds immense potential to enhance the quality and safety of the region's food supply. Agroecology, as an approach that emphasizes the ecological interactions and processes within agricultural systems, promotes sustainable farming practices that can contribute to the production of highly nutritious and pesticide-free food (Tomich et al. 2011).

One of the key benefits of agroecological practices is the prioritization of organic and natural fertilizers over synthetic chemical inputs (Leakey 2020). This approach can lead to the reduction of pesticide residues in crops and livestock products, thereby improving food safety and mitigating the risk of pesticide-related health issues (MacCarthy et al. 2021). Furthermore, the diversification of crops, including the use of traditional and locally adapted varieties, can promote dietary diversity and enhance the nutritional content of the food produced (Leakey 2020; Nyamangara et al. 2020; MacCarthy et al. 2021).

The integration of crop rotations and intercropping in agroecological systems can also contribute to improved crop quality and nutrient availability. By maintaining ecological balance and biodiversity within agricultural systems, agroecological practices can reduce the prevalence of pests and diseases, enhance natural pollination, and promote balanced ecosystems, further contributing to the production of high-quality food (Leakey 2020; Sethuraman et al. 2021). However, the adoption of agroecological practices in sub-Saharan Africa has been hindered by various challenges, such as the sub-optimal use of external inputs, fragile and poor soils, and the reliance on rainfed agriculture (Chianu and Tsujii 2004; Leakey 2020; Selladurai et al. 2020; MacCarthy et al. 2021).

References

Abawi GS, Widmer TL (2000) Impact of soil health management practices on soilborne pathogens, nematodes, and root diseases of vegetable crops. Appl Soil Ecol 15(1):37–47. https://doi.org/10.1016/s0929-1393(00)00070-6

Alemu B, Desta H, Kinati W, Mulema AA, Gizaw S, Wieland B (2019) Application of mixed methods to identify small ruminant disease priorities in Ethiopia. Front Vet Sci 6:417. https://doi.org/10.3389/fvets.2019.00417

Arriaga FJ, Guzmán J, Lowery B (2017) Conventional agricultural production systems and soil functions. In: Soil health and intensification of agroecosystems, pp 109–125. https://doi.org/10.1016/b978-0-12-805317-1.00005-1

Asrat GA (2018) Mapping soil terrain resources and descriptions of agro-ecological zone in Dawuro and Gamo Gofa zones in south-western Ethiopia. Soil Sci Environ Manag J 9(10):164–179. https://doi.org/10.5897/jssem2018.0695

Chianu JN, Tsujii H (2004) Missing links in sustainable food production in west Africa: the case of the savannas of northern Nigeria. Sustain Dev 12(4):212–222. https://doi.org/10.1002/sd.238

Cozim-Melges F, Ripoll-Bosch R, Veen GF, Oggiano P, Bianchi FJ, van der Putten WH, van Zanten HH (2024) Farming practices to enhance biodiversity across biomes: a systematic review. NPJ Biodivers 3(1):1. https://doi.org/10.1038/s44185-023-00034-2

Devenish AJ, Schmitter P, Jellason NP, Esmail N, Abdi NM, Adanu SK et al (2023) One hundred priority questions for the development of sustainable food systems in sub-Saharan Africa. Land 12(10):1879. https://doi.org/10.22541/au.169419355.58766387/v1

Dominguez-Salas P, Kauffmann D, Breyne C, Alarcon P (2019) Leveraging human nutrition through livestock interventions: perceptions, knowledge, barriers and opportunities in the Sahel. Food Secur 11:777–796. https://doi.org/10.1007/s12571-019-00957-4

Fontaine S, Abbadie L, Aubert M, Barot S, Bloor JM, Derrien D et al (2024) Plant–soil synchrony in nutrient cycles: learning from ecosystems to design sustainable agrosystems. Glob Chang Biol 30(1):e17034. https://doi.org/10.1111/gcb.17034

Gugissa DA, Abro Z, Tefera T (2022) Achieving a climate-change resilient farming system through push–pull technology: evidence from maize farming systems in Ethiopia. Sustain For 14(5):2648. https://doi.org/10.3390/su14052648

Jayne TS, Sitko NJ, Mason NM, Skole D (2018) Input subsidy programs and climate smart agriculture: current realities and future potential. Clim Smart Agric 251–273. https://doi.org/10.1007/978-3-319-61194-5_12

Koskey G, Mburu SW, Awino R, Njeru EM, Maingi JM (2021) Potential use of beneficial microorganisms for soil amelioration, phytopathogen biocontrol, and sustainable crop production in smallholder agroecosystems. Front Sustain Food Syst 5:606308. https://doi.org/10.3389/fsufs.2021.606308

Leakey RR (2020) A re-boot of tropical agriculture benefits food production, rural economies, health, social justice and the environment. Nature Food 1(5):260–265. https://doi.org/10.1038/s43016-020-0076-z

MacCarthy DS, Adam M, Freduah BS, Fosu-Mensah BY, Ampim PA, Ly M et al (2021) Climate change impact and variability on cereal productivity among smallholder farmers under future production systems in West Africa. Sustain For 13(9):5191. https://doi.org/10.3390/su13095191

Mapfumo P, Nyasimi M, Speranza CI, Wamukoya G, Kinyangi J (2017) Climate smart agriculture partnership for Africa: prospects and challenges. https://boris.unibe.ch/106758/

Niderkorn V, Jayanegara A (2021) Opportunities offered by plant bioactive compounds to improve silage quality, animal health and product quality for sustainable ruminant production: a review. Agronomy 11(1):86. https://doi.org/10.3390/agronomy11010086

Nyamangara J, Kodzwa J, Masvaya EN, Soropa G (2020) The role of synthetic fertilizers in enhancing ecosystem services in crop production systems in developing countries. In: The role of ecosystem services in sustainable food systems. Academic Press, pp 95–117. https://doi.org/10.1016/b978-0-12-816436-5.00005-6

Panel MM (2020) Meat, milk and more: policy innovations to shepherd inclusive and sustainable livestock systems in Africa. Intl Food Policy Res Inst. https://doi.org/10.2499/9780896293861

Seiber JN, Kleinschmidt L (2014) Sustainability in foods and food production: the roles of peer reviewed science and corporate initiatives. In: Chemistry of food, food supplements, and food contact materials: from production to plate. American Chemical Society, pp 5–13. https://doi.org/10.1021/bk-2014-1159.ch00

Selladurai R, Dotaniya M, Mounisamy VC, Sinha NK, Singh VK, Chauhan SS, Tripathi AK, Kundu S (2020) Sustainable soil quality management for enhancing crop productivity in tribal areas of central India: a case study. Int J Chem Stud 8(4):983–989. https://doi.org/10.22271/chemi.2020.v8.i4g.9730

Sethuraman G, Mohd Zain NA, Yusoff S, Ng YM, Baisakh N, Cheng A (2021) Revamping ecosystem services through agroecology—the case of cereals. Agriculture 11(3):204. https://doi.org/10.3390/agriculture11030204

Shah F, Wu W (2019) Soil and crop management strategies to ensure higher crop productivity within sustainable environments. Sustain For 11(5):1485. https://doi.org/10.3390/su11051485

Tenywa MM, Bekunda M (2009) Managing soils in Sub-Saharan Africa: challenges and opportunities for soil and water conservation. J Soil Water Conserv 64(1):44A–48A. https://doi.org/10.2489/jswc.64.1.44a

Tomich TP, Brodt S, Ferris H, Galt R, Horwath WR, Kebreab E et al (2011) Agroecology: a review from a global-change perspective. Annu Rev Environ Resour 36:193–222. https://doi.org/10.1146/annurev-environ-012110-121302

Topp GC, Reynolds WD, Cook FJ, Kirby JM, Carter MR (1997) Physical attributes of soil quality. In: Developments in soil science, vol 25. Elsevier, pp 21–58. https://doi.org/10.1016/s0166-2481(97)80029-3

Chapter 11
Post-Harvest Handling Technologies

Agroecology, a holistic approach to sustainable agriculture, emphasizes the importance of post-harvest handling technologies in maintaining the quality and quantity of agricultural produce. Effective post-harvest handling can contribute to food security by reducing losses, improving the accessibility of food, and ensuring the safety and nutritional value of the products.

Various post-harvest technologies have been explored and implemented in different regions, particularly in Southeast Asia where tropical commodities pose unique challenges. These technologies range from traditional methods to more advanced techniques, each with their own advantages and considerations.

One of the key aspects of post-harvest handling is the ability to preserve the quality and freshness of perishable crops. Processing techniques such as drying, heat treatment, and fermentation can help extend the shelf life of produce and reduce microbial load, making the food more accessible and safer for consumption (Cole et al. 2018). Additionally, the use of commodity storage, an ancient practice, can significantly reduce post-harvest losses and improve food security for farmers and consumers (Olorunfemi and Kayode 2021).

Emerging technologies, such as the use of brassinosteroids, methyl jasmonates, and cold plasma techniques, are also being explored to enhance the post-harvest quality retention of fruits and other produce. These eco-friendly approaches are gaining traction as consumers and the industry move away from synthetic treatments, prioritizing sustainable and environmentally friendly solutions.

The optimization of the distribution and logistics of food flows, including improved access to transportation, infrastructure, and cold chain management, is also crucial in ensuring the sustainable supply of food, particularly as the demand for agricultural products grows in urban areas and megacities (Bautista 1990; Cole et al. 2018; Cole et al. 2018; Meena and Choudhary 2019; Olorunfemi and Kayode 2021).

M. Otieno, *Sustainable Agroecological Practices in Sub-Saharan Africa in the Face of Climate Change*, Advances in Geographical and Environmental Sciences, https://doi.org/10.1007/978-3-031-70472-7_11

In conclusion, the integration of post-harvest handling technologies into agro-ecological practices is essential for maintaining food security, reducing waste, and promoting sustainable agriculture (Bautista 1990; Cole et al. 2018; Meena and Choudhary 2019; Olorunfemi and Kayode 2021; Pace and Cefola 2021).

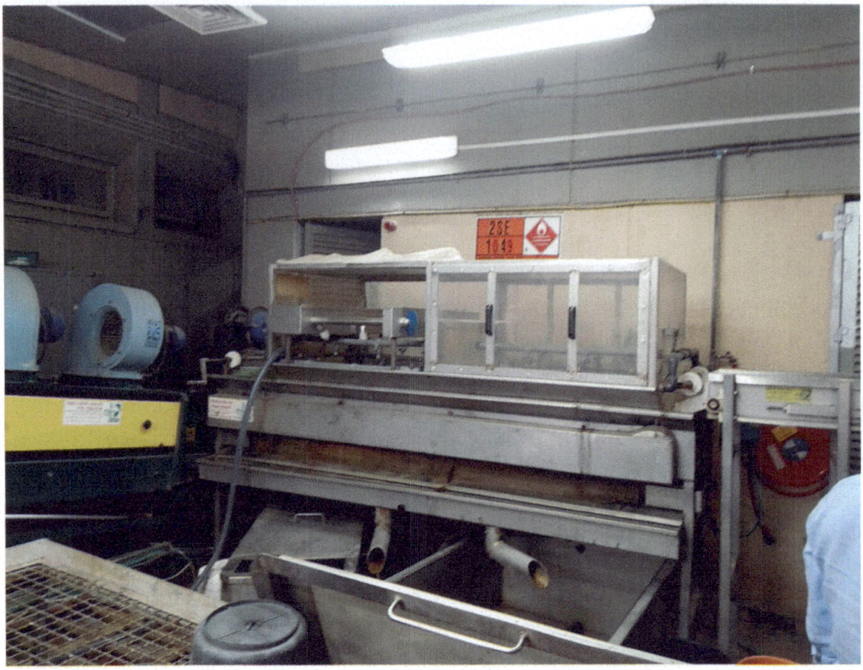

A post-Harvest handling facility

11.1 Integrating Post-Harvest Technologies and Agroecological Principles for Sustainable Food Systems in Sub-Saharan Africa

Post-harvest handling technologies play a crucial role in preserving food quality and reducing losses in sub-Saharan Africa. Effective post-harvest management is essential to minimize spoilage, maintain nutritional value, and extend the shelf life of agricultural products. When integrated with agroecological principles, these technologies can contribute to sustainable and environmentally friendly food systems.

Sub-Saharan Africa faces significant post-harvest losses due to inadequate infrastructure, limited access to storage facilities, and poor handling practices (Ibrahim et al. 2022). These losses not only reduce food availability but also contribute to economic losses for farmers. Implementing appropriate post-harvest handling technologies is crucial to mitigate these challenges and improve food security in the region (Cole et al. 2018).

Agroecological principles can guide the selection and implementation of post-harvest technologies that align with sustainable practices. For instance, the use of environmentally friendly packaging materials, such as bio-based films or natural fibers, can replace non-biodegradable plastics, reducing environmental pollution (Meena and Choudhary 2019). Agroecology emphasizes the reduction of waste and the adoption of resource-efficient practices, which can be integrated into post-harvest handling technologies to minimize the environmental footprint.

Incorporating energy-efficient technologies, such as solar dryers or improved biomass cookstoves, can also align with agroecological principles while addressing post-harvest challenges. Solar drying helps preserve the nutritional quality of crops while reducing the reliance on fossil fuels or traditional biomass, contributing to sustainable energy use. Improved cookstoves reduce smoke emissions and fuel-wood consumption during food processing, promoting environmental sustainability and reducing indoor air pollution (Bautista 1990; Cole et al. 2018; Meena and Choudhary 2019; Olorunfemi and Kayode 2021; Ibrahim et al. 2022).

Furthermore, agroecological practices can contribute to the overall resilience of food systems by promoting biodiversity, soil health, and water conservation. These practices can enhance the ability of crops to withstand post-harvest stresses, such as pests, diseases, and environmental fluctuations, thereby improving the quality and shelf life of agricultural products (Cole et al. 2018; Meena and Choudhary 2019; Ibrahim et al. 2022; Olorunfemi and Kayode 2021).

By integrating post-harvest handling technologies with agroecological principles, sub-Saharan Africa can develop sustainable and resilient food systems that address the challenges of post-harvest losses, food security, and environmental sustainability (Toivonen et al. 2014; Cole et al. 2018; Olorunfemi and Kayode 2021; Ibrahim et al. 2022).

11.2 Maintaining Food Quality Through Agroecological Practices in Sub-Saharan Africa

The use of agroecological practices in post-harvest handling is crucial for ensuring safe and nutritious food for local populations in sub-Saharan Africa. Agroecology, with its focus on integrating ecological principles into agricultural systems, extends beyond the field and emphasizes sustainable and environmentally friendly practices throughout the entire food system. Post-harvest handling, which encompasses the activities and processes involved in the handling, storage, and transportation of agricultural products after they have been harvested, is a critical juncture where agroecological practices can play a significant role in maintaining food quality.

Inadequate post-harvest handling practices in sub-Saharan Africa can lead to significant food losses, reduced food quality, and increased risks of foodborne illnesses (Ibrahim et al. 2022). Agroecological practices can contribute to addressing these challenges by prioritizing sustainable and eco-friendly techniques. For instance, the use of proper storage facilities, such as well-ventilated and pest-resistant structures, can prevent spoilage and reduce losses due to pests and diseases (Bautista 1990; Cole et al. 2018). Maintaining proper temperature and humidity conditions during storage is also essential to prevent mold growth and preserve the nutritional value of crops (Cole et al. 2018).

Agroecology also promotes the use of natural and non-chemical pest management strategies during post-harvest handling. The implementation of biological control methods, such as the use of predatory insects or microbial agents, can effectively control pests without relying on synthetic pesticides, reducing chemical residues in food and minimizing potential health risks (Meena and Choudhary 2019). Furthermore, agroecological practices emphasize the reduction of waste and the efficient use of resources (Cole et al. 2018). This includes minimizing losses during handling and transportation by employing appropriate packaging and handling techniques (Cole et al. 2018; Meena and Choudhary 2019).

11.3 Agroecological Approaches to Post-Harvest Food Safety in Sub-Saharan Africa

Ensuring food safety during post-harvest handling is crucial for the well-being and health of populations in sub-Saharan Africa. Agroecology, an approach that integrates ecological principles into agricultural systems, offers promising strategies to protect food safety throughout the entire food system, including post-harvest handling (Ibrahim et al. 2022).

Post-harvest handling encompasses the activities and processes involved in handling, storing, and transporting agricultural products after harvest. Inadequate post-harvest handling practices can lead to food contamination, spoilage, and the growth of harmful microorganisms, posing significant risks to food safety and public health. Agroecological principles can contribute to addressing these challenges in sub-Saharan Africa.

One key aspect is the use of natural and non-chemical methods for pest and disease control. Agroecology promotes the implementation of biological control strategies, such as the use of beneficial insects or microbial agents, to manage pests and diseases without relying on synthetic pesticides (Villers 2017). This approach reduces the presence of chemical residues in food, minimizing health risks associated with pesticide exposure.

Maintaining proper hygiene and sanitation practices throughout the post-harvest handling process is another important aspect of agroecological approaches. Agroecology emphasizes the use of clean and hygienic facilities, equipment, and packaging materials to prevent cross-contamination and the spread of pathogens. Proper cleaning and sanitation protocols, including regular washing and disinfection of surfaces, containers, and tools, help reduce the risk of foodborne illnesses (Villers 2017; Cole et al. 2018).

Furthermore, agroecological practices promote the reduction of post-harvest losses and waste. By implementing efficient handling and storage techniques, such as proper temperature and humidity control, packaging, and transportation, agroecological approaches can minimize spoilage and maintain the quality and safety of agricultural products (Cole et al. 2018).

Adopting agroecological principles in post-harvest handling can thus play a crucial role in protecting food safety and ensuring the well-being of populations in sub-Saharan Africa.

11.4 Reducing Post-Harvest Losses Through Agroecological Practices in Sub-Saharan Africa

Agroecological practices play a crucial role in mitigating post-harvest losses in sub-Saharan Africa, which is essential for improving food security, minimizing waste, and ensuring the efficient utilization of agricultural resources (Chianu and Tsujii 2004). Agroecology, as an approach that integrates ecological principles into agricultural systems, focuses on sustainable and environmentally friendly practices that promote the preservation and efficient use of harvested crops. Post-harvest losses in sub-Saharan Africa pose a significant challenge to food security and economic stability, as poor handling practices, inadequate storage facilities, and lack of access to markets result in substantial losses of agricultural produce (Ibrahim et al. 2022).

Agroecological practices offer solutions to address these challenges and reduce post-harvest losses. One key aspect is the promotion of appropriate storage and preservation techniques, including the use of improved storage facilities with optimal temperature, humidity, and ventilation to prevent spoilage and minimize losses, as well as the use of natural and non-chemical methods for pest control during storage. Furthermore, agroecology encourages the adoption of value-added processing techniques, such as drying, fermentation, and canning, to enhance the shelf life and market value of agricultural products (Cole et al. 2018).

Agroecological approaches prioritize the use of traditional and locally adapted processing techniques, which are often low cost and accessible to smallholder farmers (Cole et al. 2018). Post-harvest losses in the developing world, including sub-Saharan Africa, are particularly high due to factors such as improper handling, lack of protection from water loss, and inadequate transportation and storage infrastructure (Toivonen et al. 2014; Ibrahim et al. 2022). The issue is further exacerbated by the inefficient post-harvest agricultural systems in many developing countries, leading to significant food wastage and the underutilization of valuable agricultural resources. In contrast, developed countries have more extensive and effective cold chain systems to prolong product shelf life, as well as more sophisticated management and new technologies to improve the efficiency of food distribution and sale (Toivonen et al. 2014).

Addressing post-harvest losses through agroecological practices is crucial to enhance food security and ensure the sustainable use of agricultural resources in sub-Saharan Africa. Agroecological approaches offer holistic solutions that integrate ecological principles, promote appropriate storage and preservation techniques, and encourage value-added processing, ultimately contributing to the reduction of post-harvest losses and the efficient utilization of agricultural produce.

References

Bautista OK (1990) Postharvest technology for Southeast Asian perishable crops. https://www.ukdr.uplb.edu.ph/books/18/

Chianu JN, Tsujii H (2004) Missing links in sustainable food production in west Africa: the case of the savannas of northern Nigeria. Sustain Dev 12(4):212–222. https://doi.org/10.1002/sd.238

Cole MB, Augustin MA, Robertson MJ, Manners JM (2018) The science of food security. NPJ Sci Food 2(1):14. https://doi.org/10.1038/s41538-018-0021-9

Ibrahim HI, Ibrahim HY, Adeola SS, Ojoko EA (2022) Post-harvest loss and food security: a case study of major food crops in Katsina State, Nigeria. FUDMA J Agric Agric Technol 8(1):393–403. https://doi.org/10.33003/jaat.2022.0801.106

Meena NK, Choudhary K (2019) Harnessing the recent approaches in postharvest quality retention of fruits. In: Modern fruit industry. IntechOpen. https://doi.org/10.5772/intechopen.86889

Olorunfemi BJ, Kayode SE (2021) Post-harvest loss and grain storage technology – a review. Turkish J Agric Food Sci Technol 9(1):75–83. https://doi.org/10.24925/turjaf.v9i1.75-83.3714

Pace B, Cefola M (2021) Innovative preservation technology for the fresh fruit and vegetables. Food Secur 10(4):719. https://doi.org/10.3390/foods10040719

Toivonen PM, Mitcham EJ, Terry LA (2014) Postharvest care and the treatment of fruits and vegetables. In: Horticulture: plants for people and places, volume 1: production horticulture, pp 465–483. https://doi.org/10.1007/978-94-017-8578-5_13

Villers P (2017) Food safety and aflatoxin control. Canadian Center Sci Educ J Food Res 6(2):38–38. https://doi.org/10.5539/jfr.v6n2p38

Chapter 12
Climate Change Mitigation Strategies and Carbon Storage in Agricultural Systems

12.1 Climate Change Mitigation Strategies: The Role of Agroecology

Climate change mitigation strategies and agroecology are closely interconnected, as agroecological practices can contribute to reducing greenhouse gas (GHG) emissions and increasing resilience in agricultural systems. Agroecology, an approach that integrates ecological principles into agricultural systems, promotes sustainable and environmentally friendly practices that align with climate change mitigation goals.

A factory emitting greenhouse gases

One key aspect of agroecological practices that contributes to climate change mitigation is the promotion of sustainable soil management techniques. Agroecology

M. Otieno, *Sustainable Agroecological Practices in Sub-Saharan Africa in the Face of Climate Change*, Advances in Geographical and Environmental Sciences, https://doi.org/10.1007/978-3-031-70472-7_12

emphasizes practices such as organic soil amendments, cover cropping, and crop rotation, which improve soil health and increase carbon sequestration (Morgan et al. 2010; Mesfin and Markos 2019). These practices enhance soil organic matter content, leading to the retention of carbon dioxide from the atmosphere and mitigating its release as a greenhouse gas (Morgan et al. 2010; Mesfin and Markos 2019).

Furthermore, agroecology encourages the reduction of synthetic inputs, such as synthetic fertilizers and pesticides, which are energy-intensive to produce and contribute to greenhouse gas emissions. By minimizing the use of these inputs and promoting natural pest control methods and organic fertilizers, agroecological practices help reduce the carbon footprint associated with agricultural production. Agroecology also emphasizes the importance of diversifying cropping systems and promoting agrobiodiversity. By increasing crop diversity and incorporating perennial crops, agroforestry systems, and agroecosystem diversification, farmers can enhance ecosystem services and resilience in the face of climate change.

Diverse agroecosystems provide habitats for beneficial insects, promote pollination, improve water retention, and reduce the vulnerability to extreme weather events. Agroecological practices also prioritize water conservation and efficient irrigation methods, which are crucial in the face of changing climate patterns (Altieri et al. 2015).

12.2 Agroecological Practices and Carbon Sequestration: Mitigating Climate Change Through Sustainable Agriculture

Agroecological practices, such as agroforestry, cover cropping, and conservation agriculture, play a vital role in sequestering carbon dioxide (CO_2). from the atmosphere into soils and biomass, thereby mitigating the impacts of climate change (Morgan et al. 2010; Abbas et al. 2017).

Agroforestry systems, which integrate trees with crops or livestock, have been found to enhance carbon sequestration compared to conventional agricultural systems. Trees capture and store carbon through photosynthesis, and their roots contribute to the buildup of organic matter in the soil. Additionally, agroforestry systems provide shade, reduce soil erosion, and improve water infiltration, creating favorable conditions for carbon sequestration (Abbas et al. 2017).

Cover cropping, another agroecological practice, involves planting specific plant species during fallow periods or between cash crops. These cover crops capture atmospheric carbon through photosynthesis and incorporate it into their biomass and root systems. As cover crops decompose, they release organic matter into the soil, improving its fertility, water-holding capacity, and carbon storage potential (Abbas et al. 2017; Morgan et al. 2010).

Conservation agriculture, which involves minimizing soil disturbance, maintaining soil cover, and practicing crop rotation, also contributes to carbon sequestration.

Conservation agriculture enhances soil health and helps mitigate climate change by sequestering atmospheric CO_2. It does this by reducing tillage, conserving crop residues as mulch, and diversifying crop rotations, which together promote the accumulation of organic matter and increase soil carbon stocks (Morgan et al. 2010).

The sequestration of carbon through agroecological practices is significant in the context of climate change mitigation. Increased soil organic matter content and carbon storage not only contribute to reducing greenhouse gas emissions, but also provide other environmental benefits, such as improved soil fertility, water-holding capacity, and biodiversity conservation (Morgan et al. 2010; Abbas et al. 2017; Liu et al. 2024).

However, the duration of carbon sequestration rates under these practices is a critical factor that requires further investigation. Some studies have suggested that the potential for long-term carbon storage in agroecosystems may be limited by various biophysical and socioeconomic constraints (Mayer et al. 2018; Liu et al. 2024). Nevertheless, the implementation of agroecological practices remains a promising strategy for mitigating climate change through sustainable agriculture.

12.3 The Role of Agroforestry Systems in Climate Change Mitigation and Adaptation

Agroforestry, an approach that integrates trees with agricultural crops and/or livestock systems, plays a significant role in mitigating and adapting to climate change. Trees in agroforestry systems act as carbon sinks, absorbing and storing carbon dioxide (CO_2) from the atmosphere, a crucial process in reducing greenhouse gas emissions and addressing climate change (Verchot et al. 2007; Lasco et al. 2014). The presence of trees in agroforestry systems contributes to carbon sequestration through photosynthesis and the subsequent storage of carbon in biomass and soil. As trees grow and mature, they continue to absorb CO_2, effectively reducing its concentration in the atmosphere (Morgan et al. 2010).

Beyond carbon sequestration, agroforestry systems offer a range of additional benefits. They help reduce soil erosion through the tree canopy's shading and interception of rainfall, minimizing the impact of water runoff on the soil surface (Verchot et al. 2007; Abbas et al. 2017). The roots of trees also bind soil particles, enhancing soil structure and stability, further contributing to erosion control. Agroforestry also promotes increased biodiversity by providing habitat for a wide variety of plant and animal species, enhancing ecosystem resilience and crucial ecosystem services like natural pest control, pollination, and nutrient cycling. Agroforestry systems also contribute to improved microclimate regulation, with the tree canopy providing shade and reducing the temperature and intensity of solar radiation reaching the ground (Koskey et al. 2021).

The scientific evidence suggests that agroforestry plays a significant role in both mitigating and adapting to climate change (Verchot et al. 2007; Abbas et al. 2017;

Koskey et al. 2021). Agroforestry is recognized as an afforestation activity that not only sequesters carbon dioxide into the soil but also conserves biodiversity, protects cropland, acts as a windbreak, and provides a range of valuable resources for the farming community (Abbas et al. 2017). Farmer-managed agroforestry has been responsible for significant increases in food production, tree diversity, and the greening trends in the Sahel region of Africa, demonstrating its potential to address various environmental and agricultural challenges (Abbas et al. 2017; Koskey et al. 2021).

12.4 The Role of Organic Farming in Climate Change Mitigation

Organic farming practices, characterized by the avoidance of synthetic fertilizers and pesticides, play a significant role in climate change mitigation by reducing the emissions of nitrous oxide (N_2O), a potent greenhouse gas, from agricultural soils (Scialabba and Müller-Lindenlauf 2010). Additionally, organic farming promotes soil health and fertility through the use of organic matter and natural nutrient cycling.

Synthetic fertilizers, particularly nitrogen-based fertilizers, are a major source of nitrous oxide emissions in conventional agriculture. However, organic farming practices rely on the use of organic sources of nutrients such as compost, manure, and cover crops, which release nitrogen slowly and in a form that is more readily absorbed by plants. This results in reduced nitrogen losses and subsequently lower nitrous oxide emissions (Scialabba and Müller-Lindenlauf 2010). The use of organic matter, such as compost and crop residues, in organic farming practices enhances soil health and fertility. Organic matter improves soil structure, water-holding capacity, and nutrient availability, which in turn supports plant growth and productivity. Moreover, organic matter acts as a carbon source, stimulating beneficial microbial activity in the soil and promoting natural nutrient cycling processes (Scialabba and Müller-Lindenlauf 2010).

Studies have shown that organic farming systems have lower nitrous oxide emissions compared to conventional systems (Biernat et al. 2020; Dotaniya et al. 2020). A meta-analysis of field experiments conducted globally found that organic farming practices reduced nitrous oxide emissions by 40% on average compared to conventional systems (Biernat et al. 2020). Another study conducted in Europe reported similar findings, with organic farming emitting 30–70% less nitrous oxide compared to conventional farming (de la Cruz et al. 2023). The promotion of soil health and fertility through organic farming practices also contributes to climate change mitigation, as healthy soils with high organic matter content have greater capacity for carbon sequestration, thus reducing the concentration of greenhouse gases in the atmosphere (Scialabba and Müller-Lindenlauf 2010; Stavi and Lal 2013; de la Cruz et al. 2023).

12.5 Water Management Practices for Climate Change Mitigation in Agriculture

Efficient water management practices, including drip irrigation, rainwater harvesting, and water conservation techniques, play a crucial role in mitigating the impacts of climate change in agriculture. These practices contribute to climate change mitigation by optimizing water use and reducing associated energy consumption, minimizing water runoff and erosion, and enhancing water availability for crop production. Drip irrigation is an efficient water management technique that delivers water directly to the plant roots, minimizing water losses through evaporation and reducing overall water usage (Stroosnijder et al. 2012). Compared to conventional overhead irrigation methods, drip irrigation can achieve higher water-use efficiency and reduce water consumption by up to 50%. This reduction in water use not only conserves this valuable resource but also leads to energy savings by reducing the energy required for pumping and irrigation. Rainwater harvesting is another effective practice that involves capturing and storing rainfall for later use in agricultural activities. It helps alleviate water scarcity during dry periods and reduces reliance on freshwater sources. By collecting rainwater, farmers can enhance their water supply and reduce pressure on groundwater resources, which are often overexploited in many regions. This practice not only helps cope with drought conditions but also contributes to climate change adaptation by providing an alternative water source. Water conservation techniques, such as mulching, cover cropping, and soil moisture monitoring, help improve water retention in the soil and reduce water losses through evaporation and runoff. Mulching, for example, involves applying organic or inorganic materials on the soil surface to reduce evaporation, regulate soil temperature, and suppress weed growth. Cover cropping involves growing crops or plants between main crops, which can enhance soil moisture and reduce water runoff (Stroosnijder et al. 2012; Mubiru et al. 2018; Srivastava et al. 2023).

12.6 The Environmental Benefits of Integrated Pest Management Practices

Integrated Pest Management (IPM) strategies play a critical role in promoting sustainable agriculture by prioritizing natural pest control methods and minimizing reliance on chemical pesticides. By adopting IPM practices, farmers can reduce the need for chemical pesticides, thereby mitigating the environmental impacts associated with pesticide production and application, including greenhouse gas (GHG) emissions and water contamination.

IPM approaches aim to manage pests by integrating various pest control techniques while considering the ecological, economic, and social aspects of the farming system (Pretty and Pervez Bharucha 2015). These strategies emphasize preventive measures, such as crop rotation, habitat manipulation, and the use of resistant crop varieties, to minimize pest populations naturally (Barzman et al. 2015). By focusing on prevention and natural control, IPM reduces the reliance on chemical pesticides, which are often associated with adverse environmental effects (Pretty and Pervez Bharucha 2015).

One of the primary environmental benefits of IPM is the reduction in pesticide use. Chemical pesticides, if not used judiciously, can have negative impacts on ecosystems and human health. Pesticide production and application contribute to greenhouse gas emissions through energy-intensive manufacturing processes, transportation, and field applications (Gaffney et al. 2019). By minimizing pesticide use through IPM practices, farmers can mitigate the environmental footprint associated with pesticide production and reduce the release of greenhouse gases into the atmosphere (Gaffney et al. 2019).

Furthermore, pesticide applications can lead to water contamination, as runoff and leaching can transport these chemicals into water bodies, affecting aquatic ecosystems and potentially contaminating drinking water sources. IPM practices, by reducing pesticide use and promoting natural pest control methods, help minimize water contamination risks, safeguarding water quality and protecting the environment (Juroszek and Von Tiedemann 2011; Pretty and Pervez Bharucha 2015; Cycoń et al. 2017; Gaffney et al. 2019).

12.7 Crop Diversity: A Cornerstone of Agroecological Resilience

Agroecology, as a sustainable farming approach, emphasizes the significance of crop diversity and the utilization of locally adapted and traditional crop varieties. By promoting diverse cropping systems, agroecology enhances the resilience of agricultural systems to climate change impacts, including extreme weather events and pest/disease outbreaks (Altieri et al. 2015). Moreover, crop diversity plays a crucial role in maintaining genetic resources and ensuring food security under changing climatic conditions.

Agroecological practices encourage the cultivation of a wide range of crop species, including both traditional and locally adapted varieties. This approach acknowledges the importance of preserving and utilizing the genetic diversity found in traditional crops, which have evolved and adapted to specific local environments over generations (Gill et al. 2014; Altieri et al. 2015). By cultivating diverse crops, farmers reduce their vulnerability to climate-related risks, as different crops have varying degrees of tolerance to drought, heat, pests, and diseases. In the face of changing climatic conditions, a diverse cropping system can increase the chances of at least some crops thriving, providing a buffer against crop failures (Gill et al. 2014; Altieri et al. 2015).

Diverse cropping systems also contribute to ecosystem resilience and sustainability. A greater variety of crops promotes beneficial interactions among different species, including natural pest control, nutrient cycling, and soil improvement. For instance, companion planting, which involves growing different crops together, can help deter pests through intercropping and provide mutual support in nutrient uptake and soil protection (Altieri et al. 2015). By enhancing the ecological interactions within agroecosystems, diverse cropping systems reduce the reliance on synthetic inputs and foster natural ecological processes that contribute to the overall resilience and sustainability of the farming system (Gupta 2010; Altieri et al. 2015; Jat et al. 2021; Smith et al. 2023).

The diversification of agroecosystems through polycultures, agroforestry systems, and crop-livestock mixed systems, accompanied by organic soil management, water conservation and harvesting, and general enhancement of agrobiodiversity, can provide the most durable benefits in strengthening the resilience of farmers and rural communities to climate change (Gupta 2010; Altieri et al. 2015). Understanding the agroecological features that underlie the resilience of traditional agroecosystems is an urgent matter, as they can serve as the foundation for the design of climate change-resilient farming systems.

Rotational grazing, which involves periodically moving livestock to different grazing areas to allow vegetation to recover and minimize overgrazing. Research has demonstrated that this approach can reduce methane emissions from enteric fermentation by promoting healthier and more diverse pasture vegetation, which improves the overall diet quality for livestock (Wolf et al. 2017).

Studies have found that rotational grazing can lead to methane emission reductions of up to 32% compared to traditional grazing systems (Dini et al. 2017). Another strategy for reducing livestock-related emissions is the integration of livestock with agroforestry systems, known as silvopastoral systems.

These systems, which combine trees with pasture or forage crops, offer multiple benefits for livestock production and climate change mitigation. The presence of trees provides shade for animals, reducing heat stress and improving overall animal health. Additionally, the trees act as carbon sinks, sequestering atmospheric carbon dioxide and reducing greenhouse gas emissions (Shukla et al. 2019).

Efficient feed conversion is another important aspect of livestock management that can contribute to reducing methane emissions. By improving the nutritional quality and digestibility of animal feed, livestock can efficiently convert feed into meat, milk, or eggs, thereby reducing the amount of methane produced per unit of product. Strategies such as the addition of legumes to grass-based diets and the increased use of grains have been shown to improve feed efficiency and reduce methane emissions (Dini et al. 2017; Shukla et al. 2019).

12.8 Agroecology's Embrace of Energy Efficiency and Renewable Energy

Agroecological approaches have long emphasized the crucial role of energy-efficient practices and the adoption of renewable energy sources in promoting sustainable agricultural systems and mitigating climate change. By implementing energy-efficient strategies and transitioning to renewable energy, agroecology aims to reduce reliance on fossil fuels, minimize greenhouse gas emissions, and contribute to a more sustainable and resilient agricultural sector.

One energy-efficient practice widely adopted in agroecological systems is the use of low-energy irrigation systems. These systems aim to optimize water use and minimize energy consumption, such as through the use of drip irrigation, which delivers water directly to plant roots and reduces water wastage and energy requirements compared to traditional flood irrigation methods (Sethuraman et al. 2021). Studies have shown that adopting drip irrigation can result in significant energy savings and improve water-use efficiency in agriculture.

In addition to energy-efficient irrigation, agroecology encourages the adoption of renewable energy sources in agricultural operations. Solar-powered pumps, for instance, are a renewable energy technology commonly used in agroecological systems, harnessing solar energy to pump water and reducing dependence on fossil fuel-powered pumps. By utilizing solar energy, farmers can reduce energy costs, improve access to irrigation, and contribute to a more sustainable energy transition (Perakis et al. 2017; Cervera-Gascó et al. 2020; Aziz et al. 2021).

Furthermore, agroecological approaches also promote the use of biomass-based energy systems, where agricultural residues or dedicated energy crops can be converted into bioenergy through processes like anaerobic digestion or gasification. This bioenergy can then be used for heat and power generation, replacing fossil fuel-based energy sources.

12.9 Carbon Storage in Agroecological Systems: Strategies for Climate Change Mitigation

Mitigating climate change is a pressing global challenge, and the role of agricultural systems in carbon storage has garnered significant attention. Agroecology, with its emphasis on sustainable and regenerative practices, offers promising approaches to enhance carbon sequestration in agricultural ecosystems (Mayer et al. 2018; Horton et al. 2021; Morgan et al. 2010).

One key strategy employed in agroecology is the promotion of sustainable soil management practices (Morgan et al. 2010). These practices focus on improving soil organic matter content, which serves as a reservoir for carbon sequestration. By adopting techniques such as cover cropping, conservation agriculture, and organic amendments, agroecological systems enhance the input of organic matter into the soil, leading to increased carbon sequestration (Liu et al. 2024). These practices not only contribute to climate change mitigation but also enhance soil fertility, water retention, and overall soil health.

Data suggest that agriculture in the United States has the potential to sequester a substantial amount of atmospheric carbon dioxide through improved management practices (Morgan et al. 2010). Furthermore, studies have demonstrated the potential of agroforestry systems in sequestering significant amounts of carbon, both aboveground and belowground (Mayer et al. 2018; Horton et al. 2021). Agroforestry systems involve the integration of trees with agricultural crops and/or livestock, with trees acting as carbon sinks by absorbing carbon dioxide from the atmosphere and storing it in their biomass and in the soil.

In addition to soil management and agroforestry, agroecological practices that promote crop diversification can also contribute to carbon storage. By cultivating diverse crops, farmers can extend the period of active photosynthesis and increase biomass production, leading to greater carbon sequestration in agricultural systems. The use of locally adapted and traditional crop varieties also plays a role in this process, as they are often more resilient and better suited to local environmental conditions.

Overall, the implementation of agroecological practices holds significant promise for enhancing carbon storage in agricultural systems, thereby contributing to climate change mitigation efforts.

12.10 Agroecological Practices for Soil Organic Carbon Sequestration

Soil organic carbon (SOC) plays a crucial role in maintaining the fertility, water-holding capacity, and overall ecosystem health of agricultural systems, and agroecological practices prioritize the accumulation of SOC as an essential element of sustainable agriculture. Cover cropping, crop rotation, and composting are three key agroecological practices that can enhance SOC levels.

Cover crops, such as legumes or grasses, contribute to SOC accumulation by capturing atmospheric carbon dioxide through photosynthesis and incorporating it into their above-ground biomass and below-ground root systems. When these cover crops are terminated and incorporated into the soil, the organic matter they add becomes a valuable source of carbon for SOC (Blanco-Canqui et al. 2013; Tian et al. 2024). Crop rotation, by alternating the cultivation of different crop species in a field over time, helps break pest and disease cycles, improves soil structure, and increases the diversity and quality of organic matter inputs, ultimately leading to increased SOC (Palmer et al. 2017; Emadi et al. 2020).

Composting, the process of decomposing organic materials, allows the transformation of organic residues, such as crop residues and livestock manure, into stable organic matter that can be added to the soil, enriching it with organic carbon and improving soil structure, nutrient availability, and water retention capacity (Palmer et al. 2017; Zomer et al. 2017; Emadi et al. 2020; Tian et al. 2024).

Numerous studies have highlighted the potential for SOC sequestration in croplands through the adoption of climate-smart management practices like those employed in agroecology (Lal 2011; Blanco-Canqui et al. 2013; Söderström et al. 2014; Palmer et al. 2017; Bolinder et al. 2020; Tian et al. 2024). Recognizing the vulnerability of SOC to loss due to climate change, the accumulation of SOC through agroecological practices is a key natural solution for mitigating climate change and ensuring food security (Zomer et al. 2017; Bolinder et al. 2020; Emadi et al. 2020; Tian et al. 2024).

12.11 The Role of Agroforestry Systems in Enhancing Carbon Storage

Agroecology emphasizes the integration of trees within agricultural landscapes, harnessing their benefits to complement agricultural crops and livestock systems. Trees play a crucial role in carbon storage as they actively sequester atmospheric carbon dioxide through photosynthesis. Agroforestry systems, such as alley cropping and silvopasture, provide effective means of increasing carbon storage by enhancing above-ground biomass and soil carbon levels (Pinho et al. 2012; Abbas et al. 2017; Kanzler et al. 2021).

One key agroecological practice that promotes carbon storage is alley cropping. In alley cropping systems, rows of trees or shrubs are planted alongside agricultural crops, creating a mutually beneficial interaction. The trees sequester carbon dioxide from the atmosphere and accumulate biomass, contributing to above-ground carbon storage. Additionally, the fallen leaves, twigs, and branches provide organic matter inputs, which enrich the soil with carbon and contribute to the accumulation of soil organic carbon (Abbas et al. 2017).

Silvopasture, another agroforestry practice, involves the integration of trees with livestock grazing areas. Trees provide shade and forage for livestock, while simultaneously contributing to carbon storage. The trees sequester carbon dioxide through photosynthesis and store it in their biomass, including leaves, stems, and roots. The roots, in particular, play a significant role in carbon sequestration in the soil, as their decomposition adds organic matter, enhancing soil carbon levels and promoting long-term carbon storage.

Furthermore, agroforestry systems contribute to carbon storage by improving soil carbon levels. The presence of trees enhances nutrient cycling, increases organic matter inputs, and improves soil structure, all of which support the accumulation of soil organic carbon. The leaf litter and other organic materials from trees serve as carbon inputs to the soil, promoting the formation of stable soil organic matter (Nair 2014; Abbas et al. 2017; Kanzler et al. 2021; Liu et al. 2024).

Agroforestry is a sustainable land-use system with a promising potential to sequester atmospheric carbon into soil (Abbas et al. 2017). Agroforestry practices, such as alley cropping and silvopasture, offer a win–win situation for the farming community and environmental sustainability by sequestering carbon, conserving biodiversity, and providing multiple ecosystem services (Abbas et al. 2017; Zomer et al. 2021).

12.12 Biodiversity and Carbon Storage: The Role of Agroecological Approaches

Agroecological approaches recognize the significance of biodiversity conservation and the positive ecological interactions within agroecosystems. The promotion of high biodiversity levels in agricultural landscapes can significantly contribute to increased carbon storage through various mechanisms (Thrupp 2000; Batáry et al. 2011).

One mechanism is more efficient nutrient cycling. Biodiverse agroecosystems support a wide array of organisms, including microorganisms, insects, and earthworms, that play key roles in nutrient cycling processes (Cozim-Melges et al. 2024). These organisms decompose organic matter, release nutrients, and facilitate their uptake by plants, leading to increased plant productivity and the allocation of carbon to the below-ground biomass, which contributes to carbon storage in the soil (Batáry et al. 2011).

Agroecology also emphasizes the role of biodiversity in enhancing soil structure. Diverse plant species in agroecosystems have different root structures, depths, and functions, which contribute to improved soil aggregation, porosity, and stability. The presence of diverse roots creates pathways for water infiltration and root exudates that enhance soil microbial activity and carbon storage. The interactions between plants and soil organisms, such as mycorrhizal fungi, further enhance soil structure and carbon sequestration.

Furthermore, the growth of diverse plant species with varying carbon sequestration potentials contributes to carbon storage in agroecosystems. Different plant species exhibit variations in their growth rates, root biomass, and carbon allocation patterns. Some plant species have a higher capacity to capture and store carbon in their above-ground and below-ground biomass. By promoting a diverse array of plant species, agroecological approaches can maximize the overall carbon storage potential of the agroecosystem.

These mechanisms highlight the importance of biodiversity conservation in agroecological systems for enhancing carbon storage. As agricultural land covers a significant portion of the Earth's surface, the contribution of agroecosystems to carbon sequestration is crucial for mitigating climate change (Batáry et al. 2011; Liu et al. 2024).

12.13 Agroecology and Carbon Sequestration: The Role of Reduced Input Intensity

Agroecology offers a promising approach to mitigating climate change by reducing the reliance on synthetic inputs, such as chemical fertilizers and pesticides, in agricultural systems. This minimization of input intensity contributes to climate change mitigation through two primary mechanisms: the reduction of associated carbon emissions and the enhancement of soil carbon storage (Liu et al. 2024).

The use of chemical fertilizers in conventional agriculture is an energy-intensive process that often relies on fossil fuels (Mayer et al. 2018). The manufacturing and transportation of synthetic fertilizers contribute significantly to greenhouse gas emissions, including carbon dioxide and nitrous oxide (Liu et al. 2024). By reducing the reliance on chemical fertilizers, agroecology minimizes these associated carbon emissions, thereby contributing to climate change mitigation (Scialabba and Müller-Lindenlauf 2010).

Similarly, the use of chemical pesticides in conventional agriculture has adverse environmental consequences, including the release of greenhouse gases during their production and application. Pesticides may also persist in the environment, leading to long-term environmental degradation. Agroecological approaches prioritize natural pest control methods and the enhancement of ecosystem services, such as biological pest control, which reduce the need for chemical pesticides and their associated carbon emissions.

In addition to reducing carbon emissions, agroecology promotes practices that enhance organic matter inputs in the soil. Organic matter, derived from plant residues, cover crops, and organic amendments, plays a crucial role in carbon storage. Agroecological practices, such as cover cropping, composting, and the use of organic fertilizers, increase organic matter inputs into the soil, leading to higher soil carbon content and the formation of stable carbon compounds, contributing to long-term carbon storage (Morgan et al. 2010; Scialabba and Müller-Lindenlauf 2010; Mesfin and Markos 2019; Liu et al. 2024).

The potential of agroecology to mitigate climate change through reduced input intensity is well-documented. Studies have shown that organic production systems consistently exhibit lower greenhouse gas emissions compared to conventional systems, with a significant portion of these reductions attributable to the reduced reliance on synthetic inputs (Scialabba and Müller-Lindenlauf 2010; Stavi and Lal 2013; Arora 2019; Mesfin and Markos 2019). Moreover, the increased soil organic matter and enhanced carbon sequestration associated with agroecological practices have been shown to contribute to climate change adaptation by improving soil health and resilience.

12.14 Agroecosystem Resilience: Enhancing the Carbon Storage Capacity of Agricultural Systems

Agroecological systems have the potential to significantly contribute to the long-term carbon storage capacity of agricultural systems through their emphasis on diversification, resilience, and ecosystem-based approaches. These systems adopt practices that reduce the vulnerability of agricultural systems to climate change impacts, such as extreme weather events and soil degradation, thereby helping to maintain carbon stocks and prevent carbon losses (Zomer et al. 2021).

Diversification is a key principle of agroecology. By promoting crop diversity and incorporating multiple species within agroecosystems, agroecological systems enhance their resilience to climate change. Diverse cropping systems adapt better to changing environmental conditions and withstand extreme weather events like droughts and floods more effectively. This resilience helps maintain the continuous productivity of agroecosystems and their associated carbon stocks (Altieri et al. 2015; Liu et al. 2024). Furthermore, agroecological approaches prioritize ecosystem-based practices that emphasize the utilization of ecological processes and interactions to enhance agricultural productivity and sustainability. Ecosystem-based practices, such as the promotion of beneficial soil microorganisms, the enhancement of natural pollinators, and the conservation of biodiversity, contribute to the maintenance of carbon stocks by supporting the functioning of agroecosystems and facilitating the sequestration of carbon dioxide from the atmosphere into soils and biomass (Altieri et al. 2015; Fontaine et al. 2024).

Agroecology also addresses soil degradation, a significant concern in many agricultural systems. Practices such as cover cropping, agroforestry, and organic matter addition enhance soil health, structure, and fertility, leading to increased carbon storage capacity in healthy soils with higher organic matter content. These practices prevent soil erosion, enhance water retention, and improve nutrient cycling, further contributing to the resilience and carbon storage potential of agroecosystems.

Overall, the principles and practices of agroecology have the potential to significantly enhance the long-term carbon storage capacity of agricultural systems by reducing their vulnerability to climate change impacts and promoting resilient, diverse, and ecosystem-based approaches to agriculture (Altieri et al. 2015; Teague 2018; Zomer et al. 2021; Fontaine et al. 2024; Liu et al. 2024).

12.15 Livestock Integration in Agroecological Systems: Enhancing Carbon Storage and Mitigating Methane Emissions

Livestock management practices that integrate livestock within agroecological systems, such as rotational grazing and silvopastoral systems, can make significant contributions to carbon storage and climate change mitigation (Shukla et al. 2019).

Rotational grazing, also known as managed or planned grazing, involves the systematic movement of livestock between different pasture areas, allowing for adequate rest and recovery of grazed areas (Mo et al. 2023). This promotes the regeneration of vegetation and enhances carbon sequestration in grassland soils, as the rotation of livestock stimulates the growth of deep-rooted grasses and increases the below-ground biomass (Shukla et al. 2019). Silvopastoral systems, which integrate trees with grazing land, offer additional benefits for carbon sequestration. The presence of trees in these systems captures and stores atmospheric carbon dioxide through photosynthesis, contributing to the overall carbon stock of the system. Moreover, the combination of trees and pasture can enhance soil organic matter, nutrient cycling, and soil structure, further supporting carbon storage (Shukla et al. 2019; Khalil et al. 2019; Mo et al. 2023).

In addition to promoting carbon storage, these agroecological practices can also help reduce methane emissions, a potent greenhouse gas produced by livestock through enteric fermentation and manure management. By implementing improved grazing management techniques, such as rotational grazing, livestock have access to fresh forage, leading to better digestion and reduced methane production. Additionally, the integration of trees in silvopastoral systems can provide shade and a more favorable microclimate, reducing heat stress in livestock and consequently decreasing methane emissions (Shukla et al. 2019; Niderkorn and Jayanegara 2021; Mo et al. 2023).

The implementation of these sustainable livestock management practices, grounded in agroecological principles, offers a multifaceted approach to addressing the climate change challenges associated with the livestock sector. As highlighted in various studies, these practices not only contribute to carbon storage and methane emissions reduction but also have the potential to enhance the adaptive capacity of rural communities, particularly smallholders and pastoralists (Wolf et al. 2017; Shukla et al. 2019; Niderkorn and Jayanegara 2021; Mo et al. 2023).

References

Abbas F, Hammad HM, Fahad S, Cerdà A, Rizwan M, Farhad W, Ehsan S, Bakhat HF (2017, March 9) Agroforestry: a sustainable environmental practice for carbon sequestration under the climate change scenarios—a review. Environ Sci Pollut Res 24(12):11177–11191. https://doi.org/10.1007/s11356-017-8687-0

Altieri MA, Nicholls CI, Henao A, Lana M (2015) Agroecology and the design of climate change-resilient farming systems. Agron Sustain Dev 35(3):869–890. https://doi.org/10.1007/s13593-015-0285-2

Arora NK (2019) Impact of climate change on agriculture production and its sustainable solutions. Environ Sustain 2(2):95–96. https://doi.org/10.1007/s42398-019-00078-w

Aziz N, Amin N, Mohamad MB, Saad M, Izham M, Rahman M, Mohamad A (2021) Conceptual design of a solar powered agriculture irrigation system for rural farming. J Phys Conf Ser 2051(1):012066. https://doi.org/10.1088/1742-6596/2051/1/012066

Barzman M, Bàrberi P, Birch ANE, Boonekamp P, Dachbrodt-Saaydeh S, Graf B et al (2015) Eight principles of integrated pest management. Agron Sustain Dev 35:1199–1215. https://doi.org/10.1007/s13593-015-0327-9

Batáry P, Báldi A, Kleijn D, Tscharntke T (2011) Landscape-moderated biodiversity effects of agri-environmental management: a meta-analysis. Proc R Soc B Biol Sci 278(1713):1894–1902. https://doi.org/10.1098/rspb.2010.1923

Biernat L, Taube F, Loges R, Kluß C, Reinsch T (2020) Nitrous oxide emissions and methane uptake from organic and conventionally managed arable crop rotations on farms in Northwest Germany. Sustain For 12(8):3240. https://doi.org/10.3390/su12083240

Blanco-Canqui H, Shapiro CA, Wortmann CS, Drijber RA, Mamo M, Shaver TM, Ferguson RB (2013) Soil organic carbon: the value to soil properties. J Soil Water Conserv 68(5):129A–134A. https://doi.org/10.2489/jswc.68.5.129a

Bolinder MA, Crotty F, Elsen A, Frac M, Kismányoky T, Lipiec J et al (2020) The effect of crop residues, cover crops, manures and nitrogen fertilization on soil organic carbon changes in agroecosystems: a synthesis of reviews. Mitig Adapt Strateg Glob Chang 25:929–952. https://doi.org/10.1007/s11027-020-09916-3

Cervera-Gascó J, Montero J, del Castillo A, Tarjuelo JM, Moreno MA (2020) EVASOR, an integrated model to manage complex irrigation systems energized by photovoltaic generators. Agronomy 10(3):331. https://doi.org/10.3390/agronomy10030331

Cozim-Melges F, Ripoll-Bosch R, Veen GF, Oggiano P, Bianchi FJ, van der Putten WH, van Zanten HH (2024) Farming practices to enhance biodiversity across biomes: a systematic review. NPJ Biodivers 3(1):1. https://doi.org/10.1038/s44185-023-00034-2

de la Cruz VYV, Cheng W, Tawaraya K (2023) Yield gap between organic and conventional farming systems across climate types and sub-types: a meta-analysis. Agric Syst 211:103732. https://doi.org/10.1016/j.agsy.2023.103732

Cycoń M, Mrozik A, Piotrowska-Seget Z (2017) Bioaugmentation as a strategy for the remediation of pesticide-polluted soil: a review. Chemosphere 172:52–71. https://doi.org/10.1016/j.chemosphere.2016.12.129

Dini Y, Gere JI, Cajarville C, Ciganda VS (2017) Using highly nutritious pastures to mitigate enteric methane emissions from cattle grazing systems in South America. Anim Prod Sci 58(12):2329–2334. https://doi.org/10.1071/an16803

Dotaniya CK, Yashona DS, Aher SB, Rajput PS, Doutaniya RK, Lata M, Mohbe S (2020) Crop performance and soil properties under organic nutrient management. Int J Curr Microbiol App Sci 9(4):1055–1065. https://doi.org/10.20546/ijcmas.2020.904.125

Emadi M, Taghizadeh-Mehrjardi R, Cherati A, Danesh M, Mosavi A, Scholten T (2020) Predicting and mapping of soil organic carbon using machine learning algorithms in Northern Iran. Remote Sens 12(14):2234. https://doi.org/10.3390/rs12142234

Fontaine S, Abbadie L, Aubert M, Barot S, Bloor JM, Derrien D et al (2024) Plant–soil synchrony in nutrient cycles: learning from ecosystems to design sustainable agrosystems. Glob Chang Biol 30(1):e17034. https://doi.org/10.1111/gcb.17034

Gaffney J, Bing J, Byrne PF, Cassman KG, Ciampitti I, Delmer D, Warner D (2019) Science-based intensive agriculture: sustainability, food security, and the role of technology. Glob Food Sec 23:236–244. https://doi.org/10.1016/j.gfs.2019.08.003

Gill BS, Raupp WJ, Friebe B (2014) Dual threats of imperiled native agroecosystems and climate change to world food security: genomic perspectives. J Crop Improv 28(1):88–98. https://doi.org/10.1080/15427528.2014.865411

Gupta J (2010) A history of international climate change policy. Wiley Interdiscip Rev Clim Chang 1(5):636–653. https://doi.org/10.1002/wcc.67

Horton P, Long SP, Smith P, Banwart SA, Beerling DJ (2021) Technologies to deliver food and climate security through agriculture. Nat Plants 7(3):250–255. https://doi.org/10.1038/s41477-021-00877-2

Jat SL, Suby SB, Parihar CM, Gambhir G, Kumar N, Rakshit S (2021) Microbiome for sustainable agriculture: a review with special reference to the corn production system. Arch Microbiol 203(6):2771–2793. https://doi.org/10.1007/s00203-021-02320-8

Juroszek P, Von Tiedemann A (2011) Potential strategies and future requirements for plant disease management under a changing climate. Plant Pathol 60(1):100–112. https://doi.org/10.1111/j.1365-3059.2010.02410.x

Kanzler M, Böhm C, Freese D (2021) The development of soil organic carbon under young black locust (*Robinia pseudoacacia* L.) trees at a post-mining landscape in eastern Germany. New For 52(1):47–68. https://doi.org/10.1007/s11056-020-09779-1

Khalil MI, Francaviglia R, Henry B, Klumpp K, Koncz P, Llorente M … Nerger R (2019) Strategic management of grazing grassland systems to maintain and increase organic carbon in soils. CO_2 Sequestration, 1–20. https://doi.org/10.5772/intechopen.84341

Koskey G, Mburu SW, Awino R, Njeru EM, Maingi JM (2021) Potential use of beneficial microorganisms for soil amelioration, phytopathogen biocontrol, and sustainable crop production in smallholder agroecosystems. Front Sustain Food Syst 5:606308. https://doi.org/10.3389/fsufs.2021.606308

Lal R (2011) Sequestering carbon in soils of agro-ecosystems. Food Policy 36:S33–S39. https://doi.org/10.1016/j.foodpol.2010.12.001

Lasco RD, Delfino RJP, Catacutan DC, Simelton ES, Wilson DM (2014) Climate risk adaptation by smallholder farmers: the roles of trees and agroforestry. Curr Opin Environ Sustain 6:83–88. https://doi.org/10.1016/j.cosust.2013.11.013

Liu L, Zhou W, Guan K, Peng B, Xu S, Tang J et al (2024) Knowledge-guided machine learning can improve carbon cycle quantification in agroecosystems. Nat Commun 15(1):357. https://doi.org/10.1038/s41467-023-43860-5

Mayer A, Hausfather Z, Jones AD, Silver WL (2018) The potential of agricultural land management to contribute to lower global surface temperatures. Sci Adv 4(8):eaaq0932. https://doi.org/10.1126/sciadv.aaq0932

Mesfin SG, Markos E (2019) A review on the role of climate smart agriculture for sustainable crop production. J Biol Agric Healthc 9:12. https://doi.org/10.7176/jbah/9-12-02

Mo L, Zohner CM, Reich PB, Liang J, De Miguel S, Nabuurs GJ et al (2023) Integrated global assessment of the natural forest carbon potential. Nature 624(7990):92–101. https://doi.org/10.1038/s41586-023-06723-z

Morgan JA, Follett RF, Allen LH, Del Grosso S, Derner JD, Dijkstra F et al (2010) Carbon sequestration in agricultural lands of the United States. J Soil Water Conserv 65(1):6A–13A. https://doi.org/10.2489/jswc.65.1.6a

Mubiru DN, Radeny M, Kyazze FB, Zziwa A, Lwasa J, Kinyangi J, Mungai C (2018) Climate trends, risks and coping strategies in smallholder farming systems in Uganda. Clim Risk Manag 22:4–21. https://doi.org/10.1016/j.crm.2018.08.004

Nair PKR (2014) Agroforestry: practices and systems, an introduction to agroforestry, 270–282. https://doi.org/10.1016/b978-0-444-52512-3.00021-8

Niderkorn V, Jayanegara A (2021) Opportunities offered by plant bioactive compounds to improve silage quality, animal health and product quality for sustainable ruminant production: a review. Agronomy 11(1):86. https://doi.org/10.3390/agronomy11010086

Palmer J, Thorburn PJ, Biggs JS, Dominati EJ, Probert ME, Meier EA et al (2017) Nitrogen cycling from increased soil organic carbon contributes both positively and negatively to ecosystem services in wheat agro-ecosystems. Front Plant Sci 8:731. https://doi.org/10.3389/fpls.2017.00731

Perakis C, Kyriakarakos G, Hani NB, Hammad S, Damasiotis M (2017) Investigation of solar-powered drip irrigation: the case study of the Jordan Valley-short communication. https://doi.org/10.17221/12/2016-rae

Pinho RCD, Miller RP, Alfaia SS (2012) Agroforestry and the improvement of soil fertility: a view from amazonia. Hindawi Publ Corp Appl Environ Sci 2012:1–11. https://doi.org/10.1155/2012/616383

Pretty J, Pervez Bharucha Z (2015) Integrated pest management for sustainable intensification of agriculture in Asia and Africa. Insects 6(1):152–182. https://doi.org/10.3390/insects6010152

Scialabba NEH, Müller-Lindenlauf M (2010) Organic agriculture and climate change. Renew Agric Food Syst 25(2):158–169. https://doi.org/10.1017/s1742170510000116

Sethuraman G, Mohd Zain NA, Yusoff S, Ng YM, Baisakh N, Cheng A (2021) Revamping ecosystem services through agroecology—the case of cereals. Agriculture 11(3):204. https://doi.org/10.3390/agriculture11030204

Shukla PR, Skeg J, Buendia EC, Masson-Delmotte V, Pörtner HO, Roberts DC … Malley J (2019) Climate Change and Land: an IPCC special report on climate change, desertification, land degradation, sustainable land management, food security, and greenhouse gas fluxes in terrestrial ecosystems. https://philpapers.org/rec/SHUCCA-2

Smith ME, Vico G, Costa A, Bowles T, Gaudin AC, Hallin S et al (2023) Increasing crop rotational diversity can enhance cereal yields. Commun Earth Environ 4(1):89. https://doi.org/10.1038/s43247-023-00746-0

Söderström B, Hedlund K, Jackson LE, Kätterer T, Lugato E, Thomsen IK, Bracht Jørgensen H (2014) What are the effects of agricultural management on soil organic carbon (SOC) stocks? Environ Evid 3:1–8. https://doi.org/10.1186/2047-2382-3-2

Srivastava S, Basche A, Traylor E, Roy T (2023) The efficacy of conservation practices in reducing floods and improving water quality. Front Environ Sci 11:1136989. https://doi.org/10.3389/fenvs.2023.1136989

Stavi I, Lal R (2013) Agroforestry and biochar to offset climate change: a review. Agron Sustain Dev 33:81–96. https://doi.org/10.1007/s13593-012-0081-1

Stroosnijder L, Moore D, Alharbi A, Argaman E, Biazin B, van den Elsen E (2012) Improving water use efficiency in drylands. Curr Opin Environ Sustain 4(5):497–506. https://doi.org/10.1016/j.cosust.2012.08.011

Teague WR (2018) Forages and pastures symposium: Cover crops in livestock production: whole-system approach: managing grazing to restore soil health and farm livelihoods. J Anim Sci 96(4):1519–1530. https://doi.org/10.1093/jas/skx060

Thrupp LA (2000) Linking agricultural biodiversity and food security: the valuable role of agrobiodiversity for sustainable agriculture. Int Aff 76(2):265–281. https://doi.org/10.1111/1468-2346.00133

Tian J, Dungait JA, Hou R, Deng Y, Hartley IP, Yang Y et al (2024) Microbially mediated mechanisms underlie soil carbon accrual by conservation agriculture under decade-long warming. Nat Commun 15(1):377. https://doi.org/10.1038/s41467-023-44647-4

Verchot LV, Van Noordwijk M, Kandji S, Tomich T, Ong C, Albrecht A et al (2007) Climate change: linking adaptation and mitigation through agroforestry. Mitig Adapt Strateg Glob Chang 12:901–918. https://doi.org/10.1007/s11027-007-9105-6

Wolf J, Asrar GR, West TO (2017) Revised methane emissions factors and spatially distributed annual carbon fluxes for global livestock. Carbon Balance Manag 12:1–24. https://doi.org/10.1186/s13021-017-0084-y

Zomer RJ, Bossio DA, Sommer R, Verchot LV (2017) Global sequestration potential of increased organic carbon in cropland soils. Sci Rep 7(1):15554. https://doi.org/10.1038/s41598-017-15794-8

Zomer RJ, Bossio DA, Sommer R, Verchot LV (2021) Author correction: global sequestration potential of increased organic carbon in cropland soils. Sci Rep 11:18720. https://doi.org/10.1038/s41598-021-98375-0

Chapter 13
Climate Change and Its Consequences on Future Agriculture in Sub-Saharan Africa

Climate change poses a significant threat to the future of agriculture in sub-Saharan Africa (Devenish et al. 2023). The region is particularly vulnerable to the adverse effects of rising temperatures, altered rainfall patterns, and increased frequency of extreme weather events (Ibe and Amikuzuno 2019).

One of the primary concerns is the impact of climate change on crop productivity (Ibe and Amikuzuno 2019; Quenum et al. 2021). Changes in temperature and precipitation can disrupt growing seasons, leading to reduced crop yields and increased vulnerability to pests and diseases. Rising temperatures may also result in heat stress for crops, affecting their physiological processes and reducing their productivity (Ibe and Amikuzuno 2019; Gugissa et al. 2022). These changes have significant implications for food production and the livelihoods of farmers in the region.

The livestock sector in sub-Saharan Africa is also vulnerable to climate change impacts. Rising temperatures, changing rainfall patterns, and increased frequency of droughts can affect the availability and quality of pasture, water resources, and animal health. Heat stress, water scarcity, and changes in disease patterns can lead to reduced livestock productivity, increased mortality rates, and decreased resilience of livestock systems. This, in turn, can have severe consequences for the livelihoods and food security of pastoralist communities that heavily depend on livestock rearing (Pereira 2017).

Moreover, climate change is expected to significantly impact water availability in sub-Saharan Africa. Changes in rainfall patterns can lead to shifts in the timing, duration, and intensity of precipitation, affecting the availability of water resources for agricultural purposes (Serdeczny et al. 2017; Ibe and Amikuzuno 2019). This can have far-reaching consequences for crop production, livestock systems, and overall food security in the region (Serdeczny et al. 2017; Pereira 2017; Ibe and Amikuzuno 2019; Devenish et al. 2023).

M. Otieno, *Sustainable Agroecological Practices in Sub-Saharan Africa in the Face of Climate Change*, Advances in Geographical and Environmental Sciences, https://doi.org/10.1007/978-3-031-70472-7_13

The consequences of climate change on agriculture in sub-Saharan Africa are multifaceted and interconnected. Researchers have highlighted the urgent need to identify and implement strategies that make food systems more resilient in the face of increasing climate variability (Jayne et al. 2018). These include investing in climate-smart agricultural practices, improving water management, and building the skills and capacity of rural communities to adapt to the changing climate (Serdeczny et al. 2017; Jayne et al. 2018; Ateeq-Ur-Rehman et al. 2018; Devenish et al. 2023).

13.1 The Impact of Rising Temperatures on Crop and Livestock Production

The global increase in temperatures has profound implications for agricultural systems, affecting both crop growth and development as well as livestock productivity and health. High temperatures, particularly during sensitive growth stages, can disrupt various physiological processes in plants, leading to reduced photosynthesis, impaired pollen viability, and decreased seed set, ultimately diminishing crop yields (Morton 2007; Schmidhuber and Tubiello 2007; Redden et al. 2014). Different crops have varying thresholds for heat stress, with cereal crops like maize, wheat, and rice being particularly sensitive to high temperatures during their reproductive stages (Morton 2007). Studies have shown that even moderate temperature increases can have negative impacts on the yields of these staple crops in tropical regions (Redden et al. 2014).

The impact of rising temperatures is not limited to crop production; it also has significant implications for livestock productivity and well-being. Animals, especially those raised for meat and milk production, are susceptible to heat stress, which can compromise their performance and health. High ambient temperatures increase respiration rates, leading to excessive water and electrolyte loss, decreased feed intake, reduced milk production, and weight loss in livestock. Heat stress can also impact reproductive performance, immune function, and increase susceptibility to diseases in animals.

To mitigate the negative impacts of rising temperatures on crops and livestock, a range of adaptation strategies can be implemented. These include the adoption of heat-tolerant crop varieties, adjusting planting dates, implementing shading techniques, and improving animal cooling and ventilation systems. Additionally, sustainable land management practices, such as agroforestry and conservation agriculture, can help to enhance the resilience of agricultural systems to the challenges posed by climate change (Morton 2007; Redden et al. 2014; Kumar et al. 2018a, 2018b, 2018c).

13.2 Adapting to Altered Rainfall Patterns: Strategies for Sustainable Agriculture

Climate change has had a profound impact on precipitation patterns, resulting in alterations in the timing, intensity, and distribution of rainfall (Njoloma et al. 2011; Mubiru et al. 2018). One of the significant consequences is the increased frequency and intensity of extreme weather events such as droughts and floods, which pose significant challenges for agricultural systems.

Droughts, prolonged periods of below-average rainfall, can disrupt planting and harvesting schedules, hinder crop growth, and reduce yields (Olabanji et al. 2020). In regions heavily reliant on rainfed agriculture, such as sub-Saharan Africa, irregular rainfall patterns and droughts can result in decreased crop yields and increased vulnerability to food insecurity (Calzadilla et al. 2013). The impact of droughts extends beyond crops to livestock production, as water scarcity affects the availability of drinking water and reduces the quality and quantity of forage resources, leading to decreased feed intake, weight loss, and reduced milk production in dairy animals.

To adapt to changing precipitation patterns and mitigate the impact of droughts, various strategies can be employed. These include the adoption of drought-tolerant crop varieties, improved water management techniques such as rainwater harvesting and efficient irrigation systems, and the implementation of conservation agriculture practices (Olabanji et al. 2020). Farmers in Africa have employed a range of adaptation strategies, such as using drought-resistant crop varieties, practicing crop diversification, adjusting cropping patterns and planting calendars, and implementing soil and moisture conservation techniques (Akinnagbe and Irohibe 2015). Furthermore, improving and strengthening human capital through education, outreach programs, and extension services can enhance the capacity of farmers to adapt to climate change impacts.

As climate change continues to affect precipitation patterns, the need for comprehensive and integrated approaches to sustainable agricultural adaptation becomes increasingly urgent. Investing in research, developing context-specific solutions, and empowering rural communities will be crucial in building resilient and adaptive food systems in the face of these climatic challenges (Akinnagbe and Irohibe 2015; Jayne et al. 2018; Raaijmakers and Swanepoel 2020; Lindvall et al. 2020; Devenish et al. 2023).

13.3 The Impact of Increased Frequency of Extreme Weather Events on Agricultural Systems

Climate change is closely linked to the occurrence of more frequent and intense extreme weather events, such as hurricanes, storms, and heatwaves, which pose significant threats to agricultural systems (Sivakumar 2020; Olabanji et al. 2020).

Hurricanes and storms, characterized by strong winds and heavy rainfall, can lead to the uprooting or damage of crops, soil erosion, waterlogging, and flooding, thereby reducing crop yields and quality. The destruction of agricultural infrastructure, including farm buildings, irrigation systems, and storage facilities, further exacerbates the challenges faced by farmers in the aftermath of these events.

On the other hand, heatwaves, which are prolonged periods of unusually high temperatures, can have detrimental effects on crop growth and development, particularly during sensitive stages like flowering and fruit set. High temperatures can lead to physiological stress in plants, negatively impacting photosynthesis, pollination, and overall crop yields (Olabanji et al. 2020; Kumar et al. 2023). Furthermore, heatwaves can increase the water demand from crops, exacerbating water stress in regions already experiencing water scarcity.

The consequences of extreme weather events on agricultural systems are far-reaching, with implications for food security and livelihoods. Production losses due to damaged crops, infrastructure, and reduced yields can lead to reduced food availability and increased food prices, disproportionately affecting vulnerable populations. To mitigate the impact of these events, it is crucial to adopt agrometeorological strategies, such as improving and strengthening agrometeorological networks, establishing efficient agrometeorological databases, and ensuring timely dissemination of early warnings to the farming community (Sivakumar 2020; Olabanji et al. 2020). Additionally, scientific research on climate extreme risk patterns and the development of resilient crop varieties can enhance the ability of agricultural systems to withstand these challenges (Yumurtaç 2015; Sivakumar 2020; Olabanji et al. 2020).

13.4 The Implications of Climate Change on Pest and Disease Dynamics in Sub-Saharan Africa

Sub-Saharan Africa is highly vulnerable to the impacts of climate change, which are projected to profoundly affect various natural and human systems in the region (Serdeczny et al. 2017). One critical aspect of this is the potential changes in the dynamics of pests and diseases, which can have significant implications for agriculture, food security, and public health.

Climate change is expected to drive an increase in temperature and altered rainfall patterns in sub-Saharan Africa, which can create conditions that are more favorable for the proliferation of certain pests and disease-causing organisms (Bhattacharjee et al. 2011; Serdeczny et al. 2017; Ibe and Amikuzuno 2019; Gugissa et al. 2022; Devenish et al. 2023). Warmer temperatures can accelerate the life cycles of some insect pests, leading to increased population growth and spread, while changes in rainfall can create conditions that are more suitable for the growth and transmission of pathogens (Gugissa et al. 2022). Additionally, extreme weather events like droughts and floods can disrupt natural ecosystem balances and make agricultural systems more vulnerable to pest and disease outbreaks (Gugissa et al. 2022).

The potential for increased pest and disease pressures poses a serious threat to food security in the region, as smallholder farmers often lack the resources and capacity to effectively manage these challenges. Climate change-induced changes in pest and disease dynamics could reduce crop yields and agricultural productivity, further exacerbating the region's food insecurity (Ibe and Amikuzuno 2019). Beyond the agricultural sector, changes in the distribution and prevalence of disease vectors like mosquitoes could also increase the risk of vector-borne diseases for human populations (Gugissa et al. 2022).

Building climate-resilient and sustainable farming systems will be crucial to mitigate these threats. Strategies such as the adoption of integrated pest management (IPM). practices, the development of climate-smart crop varieties, and the enhancement of community-level adaptation capacities can all help strengthen the region's ability to cope with the evolving pest and disease landscape (Serdeczny et al. 2017; Jayne et al. 2018; Gugissa et al. 2022; Devenish et al. 2023). Coordinated efforts across multiple sectors will be necessary to address the complex and far-reaching implications of climate change on pest and disease dynamics in sub-Saharan Africa.

13.5 The Impact of Climate Change on Water Availability and Irrigation in Sub-Saharan Africa

Climate change poses significant challenges to water availability and irrigation systems in sub-Saharan Africa. Changes in rainfall patterns, including shifts in the timing, intensity, and duration of precipitation events, can lead to both drought conditions and increased flooding, creating difficulties for agricultural water management (Kahinda et al. 2010; Dennis and Dennis 2012). Reduced water availability due to decreased rainfall can necessitate greater reliance on irrigation to maintain crop productivity, but climate change can also disrupt traditional irrigation practices by affecting the accessibility of water resources, such as rivers, lakes, and groundwater (Kahinda et al. 2010; Calzadilla et al. 2013; Muhoyi and Mbonigaba 2021).

Climate variability, particularly changes in rainfall patterns, is particularly problematic for rainfed agriculture, as soil moisture limitations can diminish crop yields and increase the risk of rainfed farming systems (Calzadilla et al. 2013). Although the use of irrigation can mitigate the risk of climate variability, irrigated farming systems are dependent on reliable water resources and may be exposed to changes in the spatial and temporal distribution of river flow. In drier regions of sub-Saharan Africa, where a significant portion of the population relies on rainfed agriculture, the impacts of climate change on water availability pose a significant threat to food security.

Climate change not only affects crop production, but also has implications for livestock. Limited water resources can lead to dehydration and inadequate water intake, negatively impacting livestock health, growth, and productivity. Additionally, reduced water availability can limit the availability of grazing lands, forcing livestock to rely on lower-quality forage or increasing competition for scarce water sources (Calzadilla et al. 2013).

Addressing these challenges will require a multifaceted approach, including the adoption of innovative water management strategies, such as rainwater harvesting, to enhance water security and build resilience to the impacts of climate change (Kahinda et al. 2010). Strengthening the capacity of rural communities to modernize agricultural systems and adapt to changing environmental conditions will also be crucial for ensuring the long-term sustainability of food production in sub-Saharan Africa (Devenish et al. 2023).

13.6 The Impact of Climate Change on Biodiversity and Ecosystem Services in Agriculture

Climate change poses significant risks to biodiversity and ecosystems, which play a crucial role in supporting agriculture and providing essential services. These risks include the loss of pollinators and disruptions to the natural habitats and ecosystems that contribute to the biological control of pests. The consequences of these impacts can be far-reaching, affecting crop productivity, food security, and the overall functioning of agricultural systems.

Pollinators, such as bees, butterflies, and other insects, are essential for the pollination of many crops. Climate change can disrupt the delicate balance between pollinators and their host plants, leading to mismatches in their interactions. Rising temperatures, changes in precipitation patterns, and shifts in flowering times can affect the availability and synchronization of flowering plants and pollinators, potentially resulting in reduced pollination rates, lower crop yields, and decreased crop quality (Jha et al. 2013).

Furthermore, climate change can impact ecosystems and natural habitats that support biological control mechanisms. Natural enemies, including beneficial insects and predators, play a vital role in regulating pest populations by preying on or parasitizing pests. Changes in temperature, rainfall patterns, and the availability of resources can alter the abundance, distribution, and behavior of these natural enemies, potentially disrupting the balance between pests and their natural control agents. This disruption can lead to increased pest pressure, requiring additional pest management interventions and potentially leading to yield losses (Sharma and Prabhakar 2014).

Conserving biodiversity and protecting ecosystems are crucial strategies for mitigating the risks climate change poses to agriculture. Research has shown that a better understanding of the biological processes involved in farm management practices is needed to build resilience and mitigate risk in agriculture under changing climatic conditions. Biodiversity is essential to maintain the key functions of ecosystems and provide vital ecosystem services, not only for its impact on production but also for fulfilling the diverse needs of farmers and society at large (Dixon 2012; Sharma and Prabhakar 2014; Briefly 2016; Rashid et al. 2021). Predicting changes in the geographical distribution and population dynamics of insect pests can be useful for adapting pest management strategies to mitigate the adverse effects of climate change on crop production (Sharma and Prabhakar 2014).

13.7 Adapting to Climate Change: Strategies for Resilient Agriculture

As climate change continues to pose significant challenges for agricultural systems, farmers must adopt a range of adaptive measures to ensure the sustainability and productivity of their operations. One crucial strategy is the selection and cultivation of drought-resistant crop varieties. These varieties, which exhibit improved water-use efficiency, deeper root systems, and enhanced tolerance to heat, can thrive in regions affected by more frequent and prolonged droughts (Altieri et al. 2015).

By utilizing drought-resistant crops, farmers can mitigate the risks associated with water scarcity and maintain crop productivity even in drought-prone areas (FAO 2021).

In addition to drought-resistant crops, water-saving irrigation techniques are another critical adaptation approach. Efficient irrigation practices, such as drip irrigation, micro-sprinklers, and precision irrigation, can optimize water use by delivering water directly to the root zone of plants, reducing evaporation losses and minimizing water wastage. These techniques not only conserve water resources but also improve crop water-use efficiency, allowing farmers to achieve higher yields with limited water availability.

However, the successful adoption of water-saving irrigation techniques requires farmers to have access to appropriate infrastructure and technical knowledge (Kannan and Anandhi 2020). Agricultural extension services and outreach programs play a crucial role in supporting farmers in acquiring the necessary skills and resources to implement these adaptations.

Alongside these technical adaptations, farmers also employ a range of other strategies to enhance the resilience of their agricultural systems. These include crop diversification, changes in cropping patterns and planting calendars, soil moisture conservation through appropriate tillage methods, and the integration of agroforestry practices.

The integration of these diverse adaptation strategies can help farmers mitigate the impacts of climate change and ensure the long-term sustainability of their agricultural systems.

References

Akinnagbe OM, Irohibe I (2015) Agricultural adaptation strategies to climate change impacts in Africa: a review. Bangladesh J Res 39(3):407–418

Altieri MA, Nicholls CI, Henao A, Lana M (2015) Agroecology and the design of climate change-resilient farming systems. Agron Sustain Dev 35(3):869–890. https://doi.org/10.1007/s13593-015-0285-2

Ateeq-Ur-Rehman M, Siddiqui BN, Hashmi N, Masud K, Adeel M, Khan MRA et al (2018) Climate change impact on rural livelihoods of small landholder: a case of Rajanpur. Pakistan Int J Appl Agric Sci 4(2):28

Bhattacharjee R, Ntare BR, Otoo E, Yanda PZ (2011) Regional impacts of climate change: Africa. In: Crop adaptation to climate change, pp 66–77

Briefly O (2016). Ethnoprimatology: primate conservation in the 21st century 50(4):569–578

Calzadilla A, Zhu T, Rehdanz K, Tol RS, Ringler C (2013) Economywide impacts of climate change on agriculture in sub-Saharan Africa. Ecol Econ 93:150–165. https://doi.org/10.1016/j.ecolecon.2013.05.006

Dennis I, Dennis R (2012) Climate change vulnerability index for South African aquifers. Water SA 38(3):417–426

Devenish AJ, Schmitter P, Jellason NP, Esmail N, Abdi NM, Adanu SK et al (2023) One hundred priority questions for the development of sustainable food systems in sub-Saharan Africa. Land 12(10):1879. https://doi.org/10.22541/au.169419355.58766387/v1

Dixon GR (2012) Climate change-impact on crop growth and food production, and plant pathogens. Can J Plant Pathol 34(3):362–379

FAO, R (2021) Climate-smart agriculture case studies 2021–Projects from around the world. Вилучено з. https://doi.org/10.4060/cb5359en

Gugissa DA, Abro Z, Tefera T (2022) Achieving a climate-change resilient farming system through push–pull technology: evidence from maize farming systems in Ethiopia. Sustain For 14(5):2648. https://doi.org/10.3390/su14052648

Ibe GO, Amikuzuno J (2019) Climate change in sub-Saharan Africa: a menace to agricultural productivity and ecological protection. J Appl Sci Environ Manag 23(2):329–335

Jayne TS, Sitko NJ, Mason NM, Skole D (2018) Input subsidy programs and climate smart agriculture: current realities and future potential. Clim Smart Agric 251–273. https://doi.org/10.1007/978-3-319-61194-5_12

Jha S, Burkle L, Kremen C (2013) Vulnerability of pollination ecosystem services

Kahinda JM, Taigbenu AE, Boroto RJ (2010) Domestic rainwater harvesting as an adaptation measure to climate change in South Africa. Phys Chem Earth Parts A/B/C 35(13–14):742–751

Kannan N, Anandhi A (2020) Water management for sustainable food production. Water 12(3):778

Kumar M, Chaudhary V, Naresh RK, Maurya OP, Pal SL (2018a) Does integrated sources of nutrients enhance growth, yield, quality and soil fertility of vegetable crops. Int J Curr Microb App Sc 7(6):125–155. https://doi.org/10.20546/ijcmas.2018.706.017

Kumar P, Thakur CL, Rai P, Attri K (2018b) Identification of existing agroforestry systems and socio-economic assessment in Kandaghat block of Solan district, Himachal Pradesh, India. Int J Curr Microbiol App Sci 7(4):3815–3826. https://doi.org/10.20546/ijcmas.2018.704.429

Kumar VS, Kumar RP, Madhuri SB, Krishna JNCH (2018c) Effect of microclimate alteration devices and feed additive on reproductive performance of Murrah buffaloes. Pharma Innov J 7(04):2463–2472

Kumar PK, Bellundagi A, Krishna II, Mallikarjuna MG, Thimmappa RK, Rai N et al (2023) Development of bread wheat (*Triticum aestivum* L) variety HD3411 following marker-assisted backcross breeding for drought tolerance. Front Genet 14:1046624. https://doi.org/10.3389/fgene.2023.1046624

Lindvall K, Kinsman J, Abraha A, Dalmar A, Abdullahi MF, Godefay H et al (2020) Health status and health care needs of drought-related migrants in the Horn of Africa—a qualitative investigation. Int J Environ Res Public Health 17(16):5917

Morton JF (2007) The impact of climate change on smallholder and subsistence agriculture. Proc Natl Acad Sci 104(50):19680–19685

Mubiru DN, Radeny M, Kyazze FB, Zziwa A, Lwasa J, Kinyangi J, Mungai C (2018) Climate trends, risks and coping strategies in smallholder farming systems in Uganda. Clim Risk Manag 22:4–21. https://doi.org/10.1016/j.crm.2018.08.004

Muhoyi E, Mbonigaba J (2021) Stakeholder consultations on small-scale irrigation schemes' constraints in Zimbabwe. Environ Dev Sustain 1–20

Njoloma HM, Kita I, Kitamura Y, Aoyagi S (2011) Effect of climate change on Rainfed maize production: assessment of maize production vs. a changing rainfall pattern in Malawi. J Rainwater Catchment Syst 16(2):25–37

Olabanji MF, Ndarana T, Davis N (2020) Impact of climate change on crop production and potential adaptive measures in the Olifants catchment, South Africa. Climate 9(1):6

Pereira L (2017) Climate change impacts on agriculture across Africa. Oxford Res Encycl Environ Sci https://doi.org/10.1093/acrefore/9780199389414.013.292

Quenum GMLD, Nkrumah F, Klutse NAB, Sylla MB (2021) Spatiotemporal changes in temperature and precipitation in West Africa. Part i: Analysis with the CMIP6 historical dataset. Water 13(24):3506

Raaijmakers S, Swanepoel PA (2020) Vulnerability, institutional arrangements and the adaptation choices made by farmers in the Western Cape province of South Africa. S Afr J Plant Soil 37(1):51–59

Rashid S, Mushtaq MB, Farooq I, Khan Z (2021) Climate smart crops for food security. In: The nature, causes, effects and mitigation of climate change on the environment. https://doi.org/10.5772/intechopen.99164

Redden RJ, Hatfield JL, Vara Prasad PV, Ebert AW, Yadav SS, O'Leary GJ (2014) Temperature, climate change, and global food security. In: Temperature and plant development, pp 181–202

Schmidhuber J, Tubiello FN (2007) Global food security under climate change. Proc Natl Acad Sci 104(50):19703–19708

Serdeczny O, Adams S, Baarsch F, Coumou D, Robinson A, Hare W et al (2017) Climate change impacts in Sub-Saharan Africa: from physical changes to their social repercussions. Reg Environ Chang 17:1585–1600

Sharma HC, Prabhakar CS (2014) Impact of climate change on pest management and food security. In: Integrated pest management. Academic Press, pp 23–36

Sivakumar MV (2020) Climate extremes and impacts on agriculture. Agroclimatol Linking Agric Climate 60:621–647

Yumurtaç A (2015) Utilization of wild relatives of wheat, barley, maize and oat in developing abiotic and biotic stress tolerant new varieties. Emirates J Food Agric 27(1):1. https://doi.org/10.9755/ejfa.v27i1.17852

The manufacturer's authorised representative in the EU is Springer Nature Customer Service Centre GmbH, Europaplatz 3, 69115 Heidelberg, Germany. If you have any concerns regarding our products, please contact ProductSafety@springernature.com

Printed and bound by CPI Group (UK) Ltd, Croydon, CR0 4YY
01/10/2026
02225910-0007